S0-ABC-913

About the author

Dr Stephen Nottingham is a biologist who specialises in crop protection. He was awarded his Ph.D. by the University of Cambridge in 1986 and has subsequently worked in research groups in the UK and the USA on projects aimed at developing novel insect pest control methods. From 1988 to 1993 he was part of the Aphid Biology Group at Imperial College and in 1994 spent a year as a Foreign Research Associate with the United States Department of Agriculture. He has contributed numerous articles to scientific journals. More recently, he has worked as a freelance consultant and writer.

EAT YOUR GENES
How genetically modified food is entering our diet

Stephen Nottingham

Zed Books Ltd
LONDON & NEW YORK

Eat Your Genes: How genetically modified food is entering our diet was first published by Zed Books Ltd, 7 Cynthia Street, London N1 9JF, UK and Room 400, 175 Fifth Avenue, New York, NY 10010, USA in 1998.

Distributed in the USA exclusively by St Martin's Press, Inc., 175 Fifth Avenue, New York, NY 10010, USA.

Cover designed by Andrew Corbett
Set in Monotype Baskerville and Univers by Ewan Smith
Printed and bound in the United Kingdom
by Biddles Ltd, Guildford and King's Lynn

Library of Congress Cataloging-In-Publication Data

Nottingham. Stephen. 1960–
Eat your genes : how genetically modified food is entering our diet / Stephen Nottingham.
p. cm.
Includes bibliographical references and index.
ISBN 1-85649-577-9 (hardcover). — ISBN 1-85649-578-7 (pbk.)
1. Food—Biotechnology. 2. Crops—Genetic engineering.
I. Title.
TP248.65.F66N67 1998
363.19'2–dc21 98–18260
CIP

A catalogue record for this book is available from the British Library

ISBN 1 85649 577 9 cased
ISBN 1 85649 578 7 limp

Contents

Introduction

It is claimed by environmental and consumer groups that most of the population in Britain has already consumed food produced using genetic engineering.[1] A similar situation occurs in many other industrialized countries. The produce from genetically modified crops, including soybean and maize, has been mixed with produce from unmodified crops for transport and marketing, making them effectively inseparable. Therefore, a wide range of foods potentially contained genetically modified ingredients; soya, for example, is used in around 60 per cent of processed food. These foods were not labelled and so consumers were given no choice in their purchasing decisions. In 1996, around 1.2 million hectares of transgenic crops, that is crops modified by genetic engineering, were grown in the USA. This area increased to 4.0 million hectares in 1997 and is likely to increase further, as multinational companies envisage most crops being transgenic in the near future. Meanwhile, a range of other genetically modified foods are poised to enter the market, including oils from transgenic rape and fruit modified for a longer shelf-life.

The impacts of genetically modified foods are also being felt in the Third World. Modified crops are intellectual property and subject to international patent law, which restricts the flow of genetic material and scientific information to developing countries and affects how farmers can grow crops. In addition, multilateral trade agreements are favouring industrialized nations and making it difficult for developing countries to regulate the activities of multinational companies or formulate national agricultural policies. Transgenic crops, modified to produce food ingredients that have been traditionally grown in the tropics, are also now threatening the economies of Third World countries.

This book aims to explain how and why genetically modified food suddenly became part of our diet. It will outline the factors responsible for driving these foods so rapidly onto the marketplace, and will also examine the wider implications of genetic engineering for countries worldwide. *Eat Your Genes* is aimed at the general reader who wants to understand more about the important developments occurring in food production as a result of recent advances in technology. The book also includes an extensive bibliography for those wanting to study the subject further.

The genetic improvement of cultivated plants and domesticated animals stretches back to the early days of agriculture. However, genetic engineering differs in fundamental ways from previous breeding techniques, as explained in Chapter 1, where the new technology is discussed in the context of the history of genetic improvements in agriculture. The scale of experimental releases of genetically modified crops is outlined, indicating the massive investment in research and development in this area, and advances in the production of food from transgenic micro-organisms, fish and animals are described. Chapter 2 briefly describes the science behind the technology, and the techniques used by genetic engineers. It explains what genes are, how they work, and the methods available for transferring them into crop plants. Readers who are either familiar with the science, or less interested in technological detail, may want to skip or browse this chapter before moving on to the chapters describing the agricultural uses to which this technology has been put. This chapter, however, together with the glossary, provides a reference for the scientific and technical words and concepts that occur in the book.

The genetic manipulation of the quantity and quality of milk is discussed in Chapter 3. The bovine somatotropin (BST) gene from cattle has been engineered into bacteria to produce commercial quantities of this hormone, for injection into cows to increase milk yields. BST was one of the first genetically engineered products used in agriculture, and the controversy surrounding its introduction to the market introduces a number of themes that recur throughout this book. Meanwhile, genetically modified sheep and cows have been produced which express human proteins in their milk. These proteins will be sold as therapeutic drugs, while nutritionally enriched milk will be sold as infant formula.

The nature of the genes transferred to crop plants forms the subject of the next three chapters. Herbicide resistance is the trait most commonly engineered into crop plants, and Chapter 4 examines how this is achieved. Herbicide-resistant crops promise immense weed control benefits for farmers, as crops will no longer be at risk from damage by weedkillers. However, critics argue that they will lead to increased herbicide use, with detrimental effects to the environment.

Crops resistant to insects, which promise pest control with reduced insecticide inputs, are described in Chapter 5. Genes for insecticidal proteins, isolated from bacteria and other plants, have been transferred into crop plants to achieve insect resistance. Careful management of resistance in insects to these toxins may, however, be necessary for the long-term effectiveness of these crops. Modifications to a group of viruses that attack insects (baculoviruses) to make them more effective as pest control agents are also discussed in this chapter. A range of transgenic crops, engineered for ease of processing and resistance to disease, are described in Chapter 6. Fruits and vegetables modified for changes in their biochemistry, to give them a longer shelf-life, including the Flavr Savr™ tomato, or enhanced nutritional properties are already finding their way onto the market. Transgenic crops with resistance to fungal and viral disease, nematode attack and frost damage are in development. The chapter also looks at probable future developments in transgenic crops, including drought-resistant and nitrogen-fixing crop varieties.

The ecological risks of releasing genetically modified organisms into the environment are examined in Chapter 7. Transgenic crops could themselves become invasive weeds, while the unplanned exchange of genetic material with related weed species may produce weeds with herbicide or insect resistance. This could threaten agricultural and natural habitats. The ecological risks posed by genetically modified micro-organisms are described, and the likelihood of transgene escape to the wider environment discussed.

The potential risks to human health from genetically modified crops are discussed in Chapter 8. The main concerns are allergic reactions to modified foods and the spread of antibiotic resistance to micro-organisms living in animal and human guts, due to the presence of marker genes in many transgenic crop plants. These risks are put into perspective by, for example, looking at the routine use of antibiotics in livestock feed.

Some ethical and moral concerns regarding genetically modifying food are raised in Chapter 9. Bioethics committees have been set up to tackle the social implications of genetic engineering. For example, some genes have been described as 'ethically sensitive', and their use in food production may be strictly controlled. The production of transgenic animals may be detrimental to the welfare of those animals, and this may be considered to outweigh the benefits gained from genetic modification. The chapter also briefly examines the issue of morality relating to human gene research and the patenting of life.

The patenting of plants is examined in Chapter 10. This chapter includes a discussion of intellectual property rights and their incorporation into the GATT (General Agreement on Tariffs and Trade), and its successor, the World Trade Organization (WTO). Patents are issued on genetically modified organisms, genes and processes of genetic manipulation. Patents are often broadly defined and have been awarded to cover any genetic manipulation of a particular crop plant. The issuing of patents has major implications for farmers in industrialized and developing countries. Major crop plants largely originated in the Third World, where their centres of natural genetic diversity exist. However, genetic modifications to these crops have been patented by companies in the industrialized world. The criticisms that genetic resources are being unfairly exploited, and that trade agreements are in conflict with the UN Treaty on Biodiversity, are also examined.

Chapter 11 outlines the regulatory framework, in the USA and the UK, for research and development concerning genetically modified organisms and food, while the post-harvest distribution and marketing of genetically modified crops in Europe are examined in Chapter 12. The political debate, including threats of a trade war, surrounding the mixed shipments of genetically modified and unmodified soya and maize to Europe is described. This chapter also looks at how opposition to genetically modified food has grown in Europe, and how it has affected decisions by member states to restrict the marketing and cultivation of modified maize.

The arguments for and against the mandatory labelling of genetically modified foods are summarized in Chapter 13. The chapter also describes the development of labelling legislation in Europe, and the importance of food producers, retailers, consumer pressure groups and governments in influencing labelling decisions.

Chapter 14 examines to what extent transgenic crops have fulfilled their initial promise and what impact they are having on the Third World. Multinational companies, promoting transgenic crops, stressed their importance for increasing crop yields as the world's population increases. Critics argued that transgenic crops can make little contribution to solving the problems of hunger and starvation, as these are caused by poverty, and political solutions are required. Genetically modified crops promised reductions in pesticide use, but critics pointed to the emphasis on crops engineered for herbicide resistance, with transgenic seeds and herbicides often produced by the same company. In addition, the transgenic crops released to date require high inputs of fertilizers, water and pesticides, and are not compatible with many of the current ideas about sustainable agriculture. Biotechnology and genetic engineering could have major economic impacts on the Third World, but not in ways that many of the early advocates of the technology had envisaged.

Mixed shipments of genetically modified and unmodified crops, and the lack of labelling, have meant that consumers have been denied the right to choose. Growing public unease about genetic engineering may act to slow the rapid expansion of genetically modified food into the marketplace, through the actions of governments and food retailers. Consumer and environmental groups have been effective at putting their message across to the public. However, multinationals, such as Monsanto, a US$9.2 billion organization, are now fighting back in the public relations war, and genetically modified foods are probably here to stay. Chapter 15 looks at the battle for the hearts and minds of consumers, as the public realizes that most of the food in their diet may soon contain ingredients originating from genetically modified organisms.

I wish to thank Fiona Russell for her advice and encouragement, and for pointing out useful sources of information. I also thank Christine Reeves for her thorough reading of the manuscript and her valuable comments.

Note

1. For example, Sheppard, J., 1996, *Spilling the Genes*, Genetics Forum, London, UK and Greenpeace Press Release. 'By Christmas, 60% of your Food could be Genetically Manipulated'. 16 October 1996. http://www.greenpeace.org.uk/science/ge/index.html

1. A brief history of genetic improvement in agriculture

Our ancestors first cultivated plants over ten thousand years ago. The earliest archaeological evidence of agriculture is from the Middle East, although agriculture developed independently in several regions around the world. Since that time, continuous improvements have been made to crop plants to meet the food requirements of human populations. The domestication and selective breeding of animals followed plant cultivation, providing additional food and also fertilizer for the crops. Biotechnology is the use of biological processes and living organisms to produce food. It can be traced back thousands of years – for instance, the fermentation of fruits and grains to make wine and beer and, more recently, the use of yeast in baking. Now, however, advances in these areas of agriculture and biotechnology are being built upon by the new technology of genetic engineering.

Artificial selection

When Charles Darwin sought to explain the evolution of new species by his theory of natural selection, he turned to improvements humans had made in varieties of animals and plants to illustrate the mechanism behind his theory.[1] In nature, Darwin observed that populations remain relatively constant, even though more offspring are produced than are apparently necessary. He further noted that variation occurred between individuals within a population. Predation, disease, competition and other factors acted to eliminate individuals. The survivors, those best adapted to their environment, passed on any advantageous characteristics to their offspring. Thus, over the course of time, the population changed in order to adapt to the environment. New species originated, by means of natural selection, as organisms became different from their ancestors.

The genetic improvement of cultivated plants and domesticated animals involves a process called *artificial selection*, where human interference directs the evolution of varieties. The selection pressures that were important in nature may be removed. It is, for example, costly for a plant to produce anti-herbivore defences, such as insecticidal chemicals or spines and other physical deterrents. Farmers may, therefore, artificially select against these characteristics. Genetic changes that would be deleterious in nature, on the other hand, may be deemed advantageous to man. As a result, the total genetic material (*gene pool*) of a cultivated plant or a domesticated animal will become ever more different from that of its wild ancestors. Artificial selection, because of its rigorously directed nature, proceeds at a faster rate than natural selection. Crops, in particular, have been artificially selected for a large number of specific characters. This has resulted in a great deal of variation – for example, rice has tens of thousands of known varieties.

The laws of inheritance

Although plant breeding has been practised for thousands of years, it became a scientific endeavour only at the beginning of the twentieth century, with the rediscovery of Gregor Mendel's work on inheritance. Mendel was an abbot who taught natural history in a monastery in Brünn in Moravia (now Brno in the Czech Republic). His observations on crosses between common garden peas (*Pisum sativum*) led him to formulate his laws of inheritance in 1866. The two laws became the foundation of the modern science of genetics. The *law of segregation* stated that each hereditary characteristic is controlled by two factors, which segregate and pass into separate reproductive cells. The *law of independent assortment* stated that pairs of factors segregate independently of each other when reproductive cells are formed. Mendel's 'factors of heredity' are now called genes.

Mendel arrived at these laws by looking at whether peas were round or wrinkled. Each pea plant, he inferred, carried two copies or *alleles* of the gene for pea shape, one inherited from its mother and one inherited from its father. A pure-breeding round pea plant has two alleles for roundness and a pure-breeding wrinkled pea plant has two alleles for wrinkledness. Mendel crossed a pure-breeding round pea variety with a wrinkled pea variety and counted the number of offspring with round or wrinkled seeds. Plants produced from crosses between two different varieties are called *hybrids*. The first-generation offspring (F1 hybrids) must contain one allele from each parent, that is one allele for roundness and one allele for wrinkledness. Mendel found that all these F1 hybrids were identical. This is because one allele is always *dominant* and the other *recessive*. The appearance of an organism is, therefore, not necessarily a guide to the copies of the genes it contains. The physical appearance of an organism is called its *phenotype* and a description of its genes is its *genotype*. In peas, roundness is dominant and wrinkledness is recessive; therefore all the peas were round. When the F1 hydrids are bred they produce reproductive cells (eggs or sperm), each containing only one copy of the shape gene, either roundness (R) or wrinkledness (W). When the reproductive cells of the F1 generation randomly mix (independently assort) during breeding, each offspring receives two alleles, one from each parent. Four possible combinations of alleles occur in the F2 hybrids: RR, RW, WR and WW. Because roundness is dominant, and will always be expressed if it is present, three round peas occur for every wrinkled pea. Shape is, however, only one of myriad characteristics controlled by genes in peas.[2]

As more characters are added the situation rapidly becomes very complicated. Mendel deliberately worked on a simple system, using true-breeding plants, choosing characters he knew not to be inherited in an erratic manner, and worked in an enclosed and sheltered courtyard. A more complex situation is usual in nature. More than two alleles may exist, for example, or gene *linkage* might occur between two characters. Linkage is where genes situated close to each other tend to stay together, rather than independently assort, causing characters to become linked in the offspring. The rapid build-up of complex combinations of characters produced by crossing becomes the raw material of the plant breeder.

In addition to the mixing of alleles, several other mechanisms can provide the plant breeder with genetic variation upon which to work. Sudden changes in the genetic material, which may cause the cell containing it to differ in appearance or behaviour from a normal type, are called *mutations*. The organism affected is called a mutant. Different types of mutation are recognized. Gene mutations are the most common type and consist of changes within a single gene. Other mutations can alter the total amount of genetic material in an organism's cells. This has occurred on a number of occasions during the evolution of plants, when the genetic material doubles up in preparation for cell division, but the cell then fails to divide. Mutations occur naturally at low rates, are usually deleterious and, therefore, are quickly eliminated by natural selection. However, occasionally an advantageous mutation occurs. In these cases, mutated genes are added to the gene pool. Plants with additional sets of their genetic material, for example, are often more vigorous and have been favoured by plant breeders.

The mutation rate may be increased artificially by radiation and certain (mutagenic) chemicals. Plant breeders have frequently used these techniques to generate new genetic material on which to work. Genetic engineering now greatly increases the possibilities for creating novel genetic material. For example, directed mutations, such as specific gene deletions, can produce new material for use in conventional plant-breeding programmes.

Cultivated plants can naturally gain or lose gene alleles by other mechanisms. Outbreeding crops, which cannot accept pollen from individuals with a similar genotype (brassicas, for example), can exchange genetic material with wild relatives. This has relevance for genetic engineering, as the risk of gene spread is a possibility with certain genetically modified crops. Alleles can also be lost through time and chance, rather than selection, by a process called *genetic drift*. It is more significant in smaller populations, where it can have the effect of reducing genetic variation.

The Green Revolution

The application of Mendel's laws to plant-breeding programmes led to the production of high-yielding hybrid seed varieties, which, in combination with fertilizers, resulted in dramatic increases in crop yields during the period 1950–84. The term 'Green Revolution' was coined to describe this agricultural success story, particularly as it applied to Asia. This breakthrough in plant breeding was thought to be the solution to the agricultural problems of the Third World. New hybrid rice strains, for example, gave yields two to three times higher than traditional varieties. In the developing world as a whole, wheat and rice production increased by about 75 per cent between 1965 and 1980.[3] This was of immense benefit to Third World populations. For example, the trebling of wheat production in India between 1966 and 1981 was estimated to be enough to feed 184 million additional people.[3] The use of hybrid seed varieties also enabled the production of the 17 most important crops in the USA to increase by over 242 per cent between 1940 and 1980, on an increased area of only 3 per cent.[3]

However, from 1984 onwards yields levelled off, or even declined, and it was realized that yield increases came at a price. The high-yielding crops of the Green Revolution required high inputs of agrochemicals, particularly fertilizers, to achieve their potential yields, and were therefore more expensive to grow. The aim was to increase output, with little concern about limiting inputs. They also required more water, in the form of irrigation, and the use of more farm machinery than traditional crop varieties. Massive increases in the use of fertilizer and pesticides occurred. This intensive use of agrochemicals degraded the environment and polluted water, while an overuse of pesticides caused resistance in pests. Farmers became dependent on supplies of agrochemicals. Genetic diversity was reduced as indigenous varieties were replaced by new hybrid seed. The new crops favoured large farms, and large landowners came to displace poor farmers, who could not benefit from the new seed varieties. In this way, subsistence crops gave way to cash crops. The new generation of transgenic crops, produced using genetic engineering in what may become called the 'gene revolution', is perpetuating some of these problems. However, this need not necessarily be the case.

Plant breeding and genetic engineering

Traditional plant breeding will continue to produce dramatic crop improvements. However, it is constrained by limitations in sexual compatibility, which prevents cross-fertilization between species. This limits the potential gene pool, that is the total number of genes and their different alleles, available for crop improvements. Genetic engineering extends this potential by creating new genetic material for breeders to work on. Once a foreign gene has been engineered into a variety, it can be passed into hybrids like any other gene using traditional breeding methods. Genetic engineering enables genes to cross the species barrier. Genes can, therefore, be transferred in ways that were not possible before, either by traditional plant-breeding methods or in nature.

There are two basic views concerning the relationship between traditional plant breeding and genetic engineering. There are those who see genetic engineering as simply an extension of plant-breeding methods – just another technique for creating beneficial genetic change. This view is presented in the literature of multinational companies. Then there are those who view genetic engineering as radically different from what has gone before, a special case that requires special treatment. This is the view of those who wish to see stricter legislation and controls applied to genetic engineering.

While conventional plant breeding primarily involves the shifting of different forms of the same gene (alleles), which are already present in a species' gene pool, to produce varieties that differ only in degree, genetic engineering usually involves the transfer of foreign genes, not previously present in a species' gene pool, into an organism. Although offering immense benefits, the integration of foreign genes is more likely to have unpredictable physiological or biochemical effects than changes in gene alleles. Bacterial or viral vectors, involved in many gene transfer processes, may themselves have undesirable properties. In addition, a range of other genes, responsible for promoting the correct functioning of the foreign gene, and acting as markers to identify transformed material, are transferred along with the gene for the desired character. Genes for antibiotic resistance, for example, are commonly used as markers, and carry their own set of risks. These, and other unique features associated with genetic engineering, suggest that genetically modified plants should be considered differently from traditionally produced crop varieties. The speed of progress in producing commercial genetically modified crops, and other organisms, has also been much quicker than during the green revolution. Genetic modification can achieve in years transformations that would have taken decades using traditional breeding techniques.

The scale of transgenic plant releases to the environment

The first foreign gene was successfully inserted into a plant in 1983, only 29 years after the discovery of the structure of DNA. Tobacco was the first transgenic plant, that is a plant containing a foreign gene, because of its importance as a model experimental plant. Tobacco accounted for a quarter of the experimental transgenic plant releases into the environment in 1989, as the basic techniques were established. In the twelve years to 1995, however, over sixty plant species had been genetically engineered and nearly three thousand field tests of transgenic crops had been conducted worldwide.[4]

The USA has the highest number of field trial releases, followed by France and Canada. By 1993, 32 countries had conducted field trials with transgenic crops, including Australia, New Zealand, Japan, China, Chile and Argentina.[5] The number of releases in Central and South America increased dramatically after 1991, reflecting the use of this region as a counter-season to the northern hemisphere by US-based multinational companies.[6] Fewer than 1 per cent of releases have been in Africa and the Middle East.[6]

Field releases of genetically transformed crops in Europe between 1992 and 1995 were conducted mainly in France (95 releases), Belgium (59), Great Britain (58)[7] and The Netherlands (51). Germany had fewer releases (22).[8] The most commonly released transgenic crops, during this period throughout Europe, were oilseed rape (96 releases), maize (63), sugar beet (45), potato (44) and tomato (19). The characteristics that were genetically modified in these crops included enhanced resistance to weed-killing herbicides (212 releases), metabolic changes and increased storage or shelf-life (45), resistance to viruses (37), resistance to insects (33), resistance to fungi (24), resistance to bacteria (6) and resistance to nematodes (1)[8] (see Chapters 4, 5 and 6).

Different transgenic crops, released outside Europe, have reflected the importance of various crops in different geographic regions. In the USA, the principal crops were maize, soybean and cotton, and in Canada, oilseed rape (including canola) and flax. The releases in New Zealand were of transgenic kiwi.[6] Potato has had the highest number of engineered characteristics: 36 by 1993.[6] Oilseed rape, maize and tomato have also been engineered for changes to a multitude of characters.

Despite the potential for improving many characteristics in crops, however, experimental releases worldwide have concentrated on producing plants with enhanced resistance to herbicides. This characteristic allows for more effective weed control in crops, as weed-killing herbicide sprays will not damage the crops themselves. Up to 1993 herbicide resistance trials predominated in every year in every geographic region, except in the Far East, where trials with transgenic crops resistant to viruses predominated.[6]

Crop plants represent the predominant group of transgenic organisms to enter the human food chain to date, largely as ingredients in processed foods. However, transgenic bacteria, fungi, animals and fish are also being developed for use in food production.

Biotechnology: transgenic bacteria and fungi

The use of microbes to produce food and industrial products by fermentation has been going on for hundreds of years, and dates from well before genetic processes were understood. Foods and drinks produced using microbes include beer, bread, cheese, yoghurt and soy sauce. The phrase 'new biotechnology' is often used to refer to the application of genetic engineering to the fermentation process. It is enabling further developments to be made to these basic foodstuffs – for example, by using genetically modified bacteria and yeasts to produce cheese, beer and bread with distinct properties.

Louis Pasteur (1822–95) was the first person to observe that different microbes gave rise to different by-products. This led to a scientific understanding of the fermentation process. A fermenter exploits this process and is essentially a large vat containing raw materials, with the addition of microbes that produce *enzymes*. Enzymes are proteins that promote specific chemical reactions. The introduction of genetic engineering technology has massively increased the potential of the fermenter.

Commercial genetic engineering was essentially developed in 1973 in the USA by Paul Berg and Herbert Boyer at Stanford University, and Stanley Cohen at the University of California, Berkeley, when they transferred genes into the bacterium *Escherichia coli* and developed the first genetically engineered products such as human insulin and hepatitis B vaccine.[9] They founded Genentech, the largest and most successful of the early biotechnology companies. These techniques were used to produce genetically modified bovine somatotropin (BST), a growth hormone produced naturally by cows. A cow supplied with additional BST will increase its milk yield. The BST gene was integrated into bacteria to produce commercial quantities of the hormone[10] (see Chapter 3).

Cheese-making involves the action of enzymes from microbes. Different sets of microbes are used to produce different cheeses, while the action of protease enzymes extracted from animals curdles the milk, turning it into solid curds and liquid whey. Chymosin, from calves' stomachs, is the most effective of these protease enzymes. The first attempt to produce a vegetarian cheese used enzymes from plants, but without success. The commercial production of vegetarian cheese involves integrating the gene for chymosin into a yeast (*Kluyveromyces lactis*). The end product does not contain the genetically engineered organism itself.

In the production of beer, starches are broken down to sugars by an enzyme called amylase, obtained from malted barley. The sugars are then fermented to alcohol by the action of yeast. The traditional yeast used in beer-making is *Saccharomyces cerevisiae*. However, enzymes from this yeast are inefficient at fermenting longer-chain sugar molecules called dextrins. A gene from a related yeast has been genetically engineered into *S. cerevisiae* to improve the efficiency of fermentation. The beer produced using genetic engineering is a lager with a low carbohydrate content, for the 'Lite' beer market.[9] Genetically modified yeasts have also been used in baking and bread-making.

Genetic modifications to animals and fish

Traditional animal breeding involves the same process of artificial selection as described for plants. Modifications were similarly limited by a species' gene pool, until the arrival of genetic engineering. During 1996, well over sixty thousand genetically engineered animals were born in the UK alone.[11] These animals were mainly produced for biomedical research. Much of the research effort has gone into producing laboratory animals prone to disease, which can be used as models for drug studies. In addition, transgenic cows, goats and sheep have been engineered to produce human proteins in their milk (see Chapter 3).

However, livestock breeders are producing transgenic cattle, sheep, pigs and chickens for human consumption, with faster growth rates, lower fat levels and increased disease resistance.[12] These animals may soon be finding their way to market. Fifty transgenic pigs were reportedly sold for human consumption in Australia in 1995.[13] The genes transferred to animals are usually responsible for the production of growth hormones, chemicals that stimulate growth. This enables meat to be produced more economically. Research is also under way to produce pigs and poultry that are more docile and therefore better suited to intensive rearing units, featherless chickens and even sheep who self-shear by shedding their own fleeces.[2]

The production of transgenic fish is now commonplace in laboratories around the world.[14] Although food fish do not have the same history of domestication and artificial selection as plants or animals, fish-farming is generally on the increase and transgenic fish may soon be farmed for human consumption. The principal genes of commercial interest are those for growth hormones, which increase growth rate. Growth hormones also facilitate seawater adaptation in salmonids. The most dramatic growth improvements to date have been observed in Pacific or Coho Salmon (*Oncorhynchus kisutch*), using growth hormone genes from salmonids.[15] These transgenic salmon had forty times the circulating growth hormone levels, and were up to thirty-seven times heavier, than non-transgenic controls. Trials of transgenic salmon, with integrated growth hormone genes, have been conducted in the USA, Scotland and South America.[16] However, studies have shown that unregulated over-production of growth hormone is detrimental to the health of Atlantic salmon (*Salmo salar*) and Pacific salmon. The use of growth hormone must also be accompanied by additional food.[14] Another species of commercial interest is the catfish (*Clarius* spp.), where increased growth rates may have benefits for food production in the Third World.

A number of Arctic fish synthesize small antifreeze proteins, which bind to ice crystals as they begin to form, stopping further ice formation. This results in a decrease in the freezing point of the fish's blood. Atlantic salmon have been engineered with an antifreeze protein gene isolated from the winter flounder (*Pseudopleuronectes americanus*).[14] This technique may result in fish being able to live in colder waters, and could increase their production by extending fish farming into new areas. Salmon have also been genetically engineered so that they no longer need to migrate from salt water to fresh water. Instead of returning to their native streams to spawn, these salmon can live and feed in the ocean and therefore increase their growth rate and subsequent economic value. Genes for disease resistance have also been experimentally integrated into fish.

Genetic engineering therefore shows great potential for continuing the genetic improvements made, particularly during the second half of the twentieth century, to varieties of crop plants, livestock animals, and strains of bacteria. Already a bewildering array of transgenic bacteria and plants are contributing to food production. Genetically modified bacteria manufacture drugs and food supplements, while contributing to the production of cheese and other foods. In the field, modified bacteria are used to prevent frost damage to strawberries. Fruits are engineered to have different composition and a longer shelf-life. Crops are engineered to resist pests and diseases, and to withstand herbicides, so that weed control becomes more efficient. Soon, food products from transgenic fish, cattle and poultry will also find their way onto supermarket shelves. Genetically modified foods have, therefore, quickly become part of our diet. How these genetic improvements were achieved and why they have occurred will be examined in later chapters.

Notes

1. Darwin, 1859.
2. Tudge, 1993.
3. Kung and Wu, 1993a.
4. *New Scientist*, 7 January 1995, pp. 21–5.
5. Dale et al., 1993.
6. Goy and Duesing, 1995.
7. The Department of the Environment (UK) in a press release on 16 January 1997 said that the secretary of state had granted 99 consents for trials of genetically modified crops since regulations came into force in 1993.
8. Landsmann and Shah, 1995.
9. Aldridge, 1996.
10. Wheale and McNally, 1990.
11. Dixon, 1995.
12. *New Scientist*, 14 November 1992, pp. 13–14.
13. Davidmann, 1996.
14. Iyengar et al., 1996.
15. Devlin et al., 1994.
16. *New Scientist*, 6 January 1996, p. 6.

2. What is genetic engineering?

The aim of genetic engineering is to introduce, enhance or delete particular characteristics of an organism. This is achieved by the manipulation of genes. The aim of this chapter is to explain what genes are, how they work, and how they are manipulated during genetic engineering. The emphasis will be on crop plants that have been used in genetically modified foods.

DNA

Genes are functional units of a molecule called *DNA* (deoxyribose nucleic acid). Genetic engineering, or recombinant DNA technology, usually involves the insertion of a gene or genes from one species into another species. Genetic information is contained in DNA, and an organism's total DNA is called its *genome*. In 1953, James Watson and Francis Crick showed the structure of DNA to be a double helix, consisting of two intertwining DNA strands held together by weak bonds linking the *bases*, parts of the molecule that can be of varying chemical composition.[1] Four bases occur: adenine (A), thymine (T), cytosine (C) and guanine (G). Particular *base pairs* form between the two nucleic acid strands, with adenine always pairing with thymine, and cytosine always pairing with guanine. Each sub-unit of a DNA strand, referred to as a *nucleotide*, has one base. The structure of DNA can be envisaged as a spiral staircase, with the base pairs forming the steps. The sequence of bases on the two strands is said to be *complementary*. If the sequence of bases on one strand is known, then the sequence on the other strand of the double helix is easily determined because of the specific way that the bases combine. For example, if A occurs on one strand, then it can be inferred that T occurs on the other strand. The sequence of bases along a DNA strand forms the *genetic code*. A gene is a discrete and inheritable unit, occupying a particular position on a DNA molecule, with a sequence of bases having a particular function. Genes are transferred between species, to produce transgenic organisms, by genetic engineering techniques. This is possible because the genetic code is universal, it is a 'language' shared by all life forms.

In the cells of all *eukaryotes* (i.e. organisms except bacteria), the DNA occurs mainly in paired structures called *chromosomes*, which consist of long strands of tightly wound DNA and protein. The chromosomes are located within the *nucleus*, the main control centre of the cell. Other cell structures (*organelles*), for example *mitochondria* and *chloroplasts*, responsible for energy production and photosynthesis in plants respectively, also have their own DNA. During reproduction, the DNA in the nucleus re-assorts, and the paired chromosomes divide to produce the reproductive cells (eggs and sperm), which have half the genetic material of normal cells. Reproductive cells from two individuals combine to produce the offspring. One copy or allele of each gene is contributed by each parent to the offspring. Mitochondrial and chloroplast DNA is maternally inherited, however, and does not re-assort. In bacteria (*prokaryotes*), which do not have a nucleus, the DNA is distributed around the cell in structures called *plasmids*.

Protein synthesis

Genetic engineering alters an organism's characteristics. This is possible because the genes that are manipulated direct the synthesis of *proteins*. A gene is said to be *expressed* when the protein it encodes is synthesized. Proteins consist of one, or more, long chains of *amino acids*. The most important and numerous proteins coded for by DNA are *enzymes*, which regulate all the biochemical processes within an organism, including the manipulation of DNA itself. Therefore, by modifying the action of enzymes, genetic engineers can potentially modify any biochemical reaction in an organism to produce a desired change in a character. The following is a brief description of how genes express or 'code for' proteins. Two key processes are the transfer of the genetic code from DNA to an intermediate messenger molecule (*transcription*), and the construction of proteins from the code on this messenger molecule (*translation*).

Transcription involves another nucleic acid molecule called *RNA* (ribonucleic acid). RNA differs from DNA in having a different sugar (ribose instead of deoxyribose) and the base uracil (U) instead of thymine (T). Different types of RNA occur, with different specific functions. During the process of transcription, messenger RNA (*mRNA*) is formed. DNA is a very long molecule, usually confined to a cell's nucleus. mRNA is a smaller and more mobile molecule that is able to carry the genetic code for one gene, transcribed from the DNA, out of the nucleus and through the jelly-like fluid of the cell (*cytoplasm*) to structures called *ribosomes*, where protein synthesis occurs. In order for DNA to be transcribed it unwinds to form two single DNA strands, a process that exposes the bases. mRNA is synthesized a bit at a time, with complementary mRNA bases being transcribed from the DNA bases – for example, if T occurs on DNA then A will be encoded on mRNA. The DNA double helix rewinds as transcription proceeds along a DNA strand. Transcription is halted when a complete gene is encoded on the mRNA. The process can be repeated many times. mRNA is therefore a complementary or reverse copy of the gene to be expressed.

The transcription process in plants and other eukaryotic organisms is complicated by the fact that genes are often not continuous coding sequences along a DNA molecule. Coding regions, called *exons*, are interrupted by non-coding regions, called *introns*. The intron regions are excised from the mRNA immediately after transcription, by enzymes that cut the nucleic acid strand on either side of the intron regions and then join or splice together the exon regions, to produce mRNA molecules with continuous complementary coding sequences.

The process of translating the genetic code from mRNA into protein molecules occurs on the ribosomes. The coding sequence on the mRNA determines the order in which the 20 possible amino acids, the building blocks of proteins, are synthesized. Another type of RNA, called transfer RNA (*tRNA*), forms the final link between mRNA and protein. There are 20 forms of tRNA, each of which attaches itself to a different amino acid. Amino acids are joined when one end of a tRNA molecule links with the corresponding coding sequence on the mRNA molecule. tRNAs translate all the genetic code on a mRNA molecule in this way, to produce an amino acid chain. Each mRNA molecule can be translated many times, to form thousands of protein molecules.

Jumping genes

The process of information flow from DNA, through mRNA, to protein, in one direction only, became known as the *Central Dogma*. For many years this was central to molecular biology and thought to be inviolable. However, information flow is no longer regarded as being unidirectional. A group of viruses, called the retroviruses, can reverse the flow of information from mRNA to DNA, using an enzyme called reverse transcriptase. This enzyme can assemble DNA from a mRNA template. Human immunodeficiency virus (HIV) is one such retrovirus. Reverse transcriptases, isolated from bacteria and viruses, have become important tools in genetic engineering.

It is now suspected that information can also pass from protein to protein. A group of infectious nervous diseases, the spongiform encephalopathies, pass information by a replication mechanism involving a protein, called a prion, seemingly without the involvement of nucleic acid. One of these diseases is bovine spongiform encephalopathy (BSE). As no nucleic acid is involved, high temperatures and ultraviolet radiation (UV) cannot kill the infective agent. Meat from cows infected with BSE has been implicated in a new variant of Creutzfeldt-Jakob disease (vCJD) in humans.

Genomes were once thought to be relatively stable, except for random mutations, and to pass unchanged to the next generation. The genome is now thought to be much more fluid and dynamic than this. Barbara McClintock was the first person to advance the idea of 'jumping genes', mobile elements that hop from one site on a chromosome to another.[2] The existence of these elements, which are now called *transposons*, was established in the 1970s. Transposons are common in bacteria, where they replicate themselves and can integrate at any point around the genome, where they can cause serious disruption to gene function. They are also common in plants, where they are referred to as transposable elements. Maize, for example, has several types, which act to move genes around the genome. Transposable elements from mitochondrial or chloroplast DNA, which can account for up to 25 per cent of the cell's total DNA in species such as maize, sorghum and sugar beet, can move to the nucleus, and from the nucleus to other organelles.[3]

It is now known that genes can undergo radical changes during an organism's lifetime, be subject to feedback and metabolic regulation and even be passed between species, via the action of viruses and bacteria. The action of a gene can be modified in response to the environment. A controversial study from 1988 suggested that bacteria could 'direct' which mutation was to occur, on the basis of what would be environmentally favourable to them. Environmental influences that produce inheritable changes in gene function, however, are now suspected in a range of organisms, including mammals. In mice, researchers at the Babraham Institute, England, and the Free University of Berlin have shown that, although changes in the coding sequence acquired during an animal's lifetime cannot be inherited, modifications to the way genes work, brought about by environmental influences, can be passed to offspring.[4]

Most genes code for proteins, but genes also have specific regulatory functions and control the expression of other genes. All genes can act to modify the effects of any other gene, through subtle environmental changes. The genome is now viewed as a network of interacting genes, rather than as a linear sequence of independently acting genes.[5] Genes are both more flexible and more dynamic than thought possible, even a couple of decades ago. This changing view of the genome has implications for genetic engineering.

Enzymes: the genetic engineer's toolkit

Enzymes are proteins that promote or catalyse specific chemical reactions. Cells use enzymes to maintain and copy DNA. The genetic engineer makes use of these enzymes as tools to manipulate DNA. Different enzymes are responsible for different tasks: unzipping double strands of DNA, cutting DNA at specific points, copying DNA, proof-reading DNA for errors and pasting sections of DNA into the genome.[6] Enzymes, whose names usually end in '-ase', feature prominently in any discussion of genetic engineering. They provide the tools for genetic manipulation, while their production by transferred foreign genes directs the changes in characteristics observed in transgenic organisms.

Restriction enzymes (restriction endonucleases) act to cut up foreign DNA. Bacteria naturally mobilize these enzymes to chop up the DNA of invading viruses. This restricts virus growth, hence the name restriction enzymes. If a virus successfully passes its genes into a bacterial genome, however, it will be protected against restriction enzymes and will be able to start controlling the cell. Restriction enzymes do not cut the DNA of their own cells. This is due to the action of enzymes that modify the bases and stop restriction enzymes recognizing the coding sequences they normally cut. This modification, however, does not affect base pairing.

Restriction enzymes were first isolated and identified in 1970. Several hundred have now been identified, each with a highly specific action. Each restriction enzyme recognizes a specific coding sequence and cuts between particular bases within this sequence. Cuts can be clean or staggered. Restriction enzymes that produce staggered cuts are particularly useful, as they leave several bases exposed, the so-called sticky ends, which will readily bind to complementary sequences of DNA from different sources. The joining of sticky ends, however, produces only weak bonding. Stronger bonds between DNA fragments can be produced using *ligase enzymes*, whose natural role in the cell is DNA repair. Ligases catalyse the formation of deoxyribose–phosphate bonds, the bonds that link nucleotides in nucleic acid chains. DNA produced by joining together fragments from different organisms is called *recombinant DNA*.

Cocktails of different restriction enzymes are available commercially and are routinely used in genetic manipulations. They are named after the bacteria from which they were first isolated. Commercially available restriction enzymes are obtained from *Escherichia coli* (e.g. enzyme *Eco* RI), *Haemophilus aegyptius* (e.g. *Hae* III), *Streptococcus albus* (e.g. *Sal* I), *Brevibacterium albidum* (e.g. *Bal* I) and a few other bacteria. These enzymes, and others used in commercial cocktails, all cut DNA at different coding sequences.

The processes of genetic manipulation have been likened to using a word-processor by Robert Pollack, in a book that considered DNA as a text.[7] The molecular word-processor's keyboard has five letters, corresponding to the five bases of DNA and RNA, and six function keys, corresponding to the functions of different groups of enzymes. The cut-and-paste keys in effect activate the microbially derived restriction and ligase enzymes, respectively. The search key uses probes of synthetic DNA to hybridize with fragments of sample DNA having complementary sequences, and the undo key activates reverse transcriptase, which can reproduce DNA sequences from the mRNA produced by active genes. The print key activates an enzyme called DNA polymerase, which produces one copy of a DNA sample, while the copy key uses DNA polymerase, combined with cycles of hybridization, to produce many copies of a stretch of DNA in a process called polymerase chain reaction (PCR).[8]

If the coding sequence is known for a particular gene, then that gene can be made in the laboratory. The gene machine or automated DNA synthesizer mimics the role of DNA polymerase in stringing together nucleotide blocks, but in the order dictated by a human operator rather than by a complementary DNA strand. These machines are now standard pieces of desktop equipment in molecular biology laboratories. Sequences for important genes are stored as programmes to allow synthetic genes to be manufactured quickly and easily. The DNA synthesizer can also be used to modify coding sequences to create novel proteins. This is known as protein engineering, and is likely to become of increasing importance in the future.[9] For example, a single base in a coding sequence could be changed, resulting in a single amino acid change in a protein. Amino acid changes can be made in existing proteins to give them additional properties, for instance heat resistance.

DNA synthesizers can be used to construct synthetic genes lacking the non-coding regions, or introns, commonly found in genes from eukaryotic organisms. To determine the base sequence for genes of this type, mRNA is obtained from a cell and a complementary copy of DNA (cDNA) made from it using reverse transcriptase. The mature mRNA will already have had its non-coding sequences excised, and will produce a single-stranded DNA without the intron regions that were present in the original DNA. This is important for plant or animal genes that are transferred to prokaryotes, because bacterial cells have no means of excising the non-coding regions of genes when transcribing mRNA from DNA.

Methods of gene transfer to crop plants

Genetic transformation, to produce transgenic plants, refers to the stable integration of a foreign gene into the genome of a plant regenerated from cells or cells stripped of their cell walls by enzymes (*protoplasts*). The transformations should be inheritable, and seed grown from transgenic plants should produce plants that also express the foreign gene. Genes are multiplied or cloned, and then transferred into plants within vehicles called *vectors*, usually derived from the small circular DNA structures in bacteria called *plasmids*. Restriction enzymes are used to cut the vector, allowing foreign genes to be inserted, while ligase enzymes rejoin the vector. The majority of gene transfer experiments up to the early 1990s were done using a bacterial vector to carry genes into a plant's genome. Methods of direct transfer of DNA, using microprojectiles, became increasingly popular during the 1990s. Other methods, including electroporation and sonication, where electric shocks and sound waves, respectively, are used to puncture holes in cell membrane in order to introduce foreign DNA, may be of value in certain circumstances.[10] A foreign gene cannot be transferred successfully without the correct gene regulation machinery being in place, either transferred along with the foreign gene or already in place in the organism receiving the gene.

Viral vectors and gene regulation

Viruses have many attributes that would suggest them as suitable vectors for transferring genes to crop plants.[11] The nucleic acid from viruses is directly infectious to plants and transfer can be achieved simply to rubbing a solution onto a leaf. Following inoculation, viruses are spread to every cell in the plant, which means that plants need not be regenerated from a single cell. Plant viruses also have a wide host range.

Viruses that attack bacteria are known as *bacteriophages*. A bacteriophage that attacks *E. coli* is commonly used as a cloning vector. Viral vectors have also been used to transfer genes into plants.[12] However, despite their many advantages, the disadvantages and potential risks have resulted in their use as transfer vectors being largely abandoned. Viruses are pathogenic agents that debilitate plants, viral nucleic acid does not integrate into a plant genome to produce stable transformations, and most plant viruses have single-stranded DNA or RNA, which are more difficult to manipulate than double-stranded DNA.[11]

Viruses do, however, provide the promoter genes that are used to produce high levels of foreign gene expression within transgenic plants. A region of the bacteriophage genome, used in cloning vectors, is used for this purpose. Promoter genes from cauliflower mosaic virus (CaMV) are also frequently used. CaMV genes are particularly convenient as they are from the only virus group known that has double-stranded DNA. These promoter genes express the enzymes that viruses naturally use to take over a bacteria cell's genetic machinery during an infection cycle. Genetic engineers exploit this in order to instruct the plant genome to express foreign genes. Promoter genes are placed in transfer vectors, usually derived from bacterial plasmids, along with the genes expressing desirable characteristics and selectable marker genes. The complete vector is sometimes referred to as a *vector construct.*

Bacterial vectors: the *Agrobacterium* method

The first transgenic plants produced, tobacco, petunia and cotton, were modified using *Agrobacterium tumefaciens* as a vector.[13] The soil-dwelling bacteria *Agrobacterium tumefaciens* and *Agrobacterium rhizogenes* are respectively the causal agents of crown gall disease and hairy root disease in plants. These bacteria naturally infect over one hundred plant species, causing aberrant plant growth by transferring some of their genes into the nuclear genome of plants. They are in effect natural genetic engineers. The genes responsible for the gene transfer are located on the plasmids. These are of two types, the Ti (Tumour-inducing) or Ri (Root-inducing) plasmids, which cause tumours or galls in plants.

The natural infection cycle of *Agrobacterium* begins when bacteria in soil become attracted to chemicals released by a wounded plant. The bacteria bind to plant cells in the wound region, where transfer of DNA to plant cells occurs. This DNA integrates into the plant's nuclear DNA. Only a relatively small and discrete part of the plasmid is transferred to the plant, a region called the T-DNA (Transferred DNA). Genes in the T-DNA region, during a natural infection, direct the synthesis of plant hormones and amino acid compounds that generally divert plant compounds into substances preferential to the bacteria. Another part of the plasmid, the *vir* (virulence) region, contains genes that direct the actual gene transfer process, but genes from this region are not transferred. T-DNA genes are, therefore, not themselves involved in the mechanics of the transfer process, with the result that the T-DNA region can be completely or partially removed and successful transfer to the plant genome can still occur. This is exploited by genetic engineers, who mimic the natural infectivity cycle in the laboratory using modified T-DNA regions containing foreign genes.

The gene for the character modification of interest is first cloned to increase the genetic material available for transferring. This is done within a cloning vector in a suitable bacterial host, usually *E. coli*. The vector construct, which also contains promoter and marker genes, is then incorporated into an *Agrobacterium* Ti or Ri plasmid for transfer to plant tissue. This plasmid is disabled, by deleting genes that normally lead to tumour or gall production. This modification means that the transformed plant cells give rise to normal-looking, fertile plants.[14] Three basic methods are used to obtain transformed plant tissue. Stem tissue can be wounded and inoculated with *Agrobacterium* either by injection or by painting a solution onto a cut surface; protoplasts can be formed and left for one or two days, so that cell walls begin to reform, and then *Agrobacterium* added; or tissue segments can be inoculated with a bacterial solution in a petri dish.[11] Tissue culture techniques are then used to raise large numbers of plants. The use of protoplasts is advantageous because the lack of cell walls facilitates the movement of foreign genes into cells, while plants regenerated from protoplasts have a uniform genetic make-up. However, by whatever method, only a small proportion will become stable transgenic plants. If integration occurs as intended, seeds of transformed plants should grow into plants with the engineered trait. Plants containing a foreign gene can then be used in conventional plant-breeding programmes.

The gene transfer method using *Agrobacterium* as a vector is labour intensive and has a major limitation in that *Agrobacterium* does not naturally infect monocotyledonous species, which include cereal crops, such as rice, wheat and maize, and the onion family.[15] Despite some modifications and limited success, the system is still only really effective for dicotyledonous crops, such as potato, tomato, soybean and sugar beet. The hurdle of transforming cereals was partially overcome in the late 1980s by workers at Sandoz, now part of Novartis, who transformed maize protoplasts using electroporation,[16] but direct transfer methods were soon available that overcame the difficulties with less effort and labour being expended.

Gene guns

Physical methods of gene transfer were developed in the late 1980s that did not require transfer using bacteria and could be used with equal ease on monocotyledonous and dicotyledonous plants. Chief among these physical methods was the use of particle bombardment, with methods developed independently by two research groups: the 'Biolistic' method[17] by John Sanford and his colleagues at Cornell University, USA, and the 'Accell' method[18] by Dennis McCabe and his colleagues at the Agracetus company, USA.

In the 'Biolistic' (biological ballistics) method, magnesium tungsten or gold particles are coated with DNA and literally blasted into plant cells using a gunpowder detonation in a particle gun. The particles carrying DNA are accelerated at high velocity, penetrate the cell wall, and enter intact plant cells without killing them. As the particles pass through the cell walls, genes are stripped from them and remain in the cell. Experimenters initially used epidermal cells of onion (*Allium cepa*) to show the potential for this technique.[17] As with the *Agrobacterium* method, the DNA transferred using particle bombardment techniques is in the form of a vector construct, which can also contain promoter genes and selectable marker genes. The Du Pont company has exclusive rights to use Cornell's patented Biolistic gene gun for developing commercial transgenic crop seed.

The 'Accell' method uses particle acceleration by electrical discharge to propel DNA-coated gold particles into plant material. Although the differences between the two gene guns are not great, they were sufficiently different to gain different patents in the late 1980s. The Agracetus patent was instrumental in gaining the company broad patent rights to genetically modified crops (see Chapter 10). In 1988, Agracetus became the first company to transfer foreign genes into soybeans. Monsanto developed its Roundup Ready™ herbicide-resistant soybeans in collaboration with Agracetus using this technology.

When plant tissue is transformed by particle bombardment, plants regenerated from the tissues are chimeric – that is, not all the cells contain foreign genes – because random bombardment affects only a small proportion of cells. This contrasts with plants regenerated from protoplasts using the *Agrobacterium* system, which have a uniform genetic make-up. The use of selectable markers is required to sort out and stabilize the progeny of the plants transformed by particle bombardment. The efficiency of delivery of foreign genes into intact cells is low compared to bacterial vector methods, but is improving with further developments in gene gun technology. Direct transfer by microprojectile methods, however, involves minimal manipulation of target tissue and is versatile, efficient and flexible. It can be used to transform any plant species and virtually any type of plant cell or tissue. It therefore allows recovery of transformed plants from tissue that could not be transformed using *Agrobacterium* or other methods.[14] Direct transfer is likely to become an increasingly favoured technique of gene transfer for these reasons.

Transgenic animals and fish are produced by another physical or direct insertion method: micro-injection. A fertilized egg is taken from an animal and is injected with the foreign DNA using a small syringe.[19] The injected DNA integrates itself randomly into the chromosomes. Many foreign genes can become integrated in this way, although in the majority of cases no foreign genes successfully integrate into the cell's genome (see Chapter 3). Fish offer a number of advantages for genetic manipulation. They have a high fecundity and external fertilization and development (unlike animal eggs, which need to be removed from the body before foreign genes can be inserted), and many species have transparent embryos. Micro-injection has also had limited success as a technique with plants, although tough cell walls make the technique difficult. The fact that any plant cell is able to regenerate into a whole plant, however, means that more techniques are potentially available for producing transgenic plants than for producing transgenic animals.

Gene silencing

Not all genetic manipulation involves transgenes that express protein. Another method is to silence an organism's own genes to prevent them being expressed. Gene-silencing manipulations involve the down regulation or suppression of genes, using antisense or sense gene constructs to block protein synthesis.[20] Gene silencing works by either preventing mRNA being formed or disabling it before it can arrive at the ribosome, the site of protein synthesis.

A sense gene has the same coding sequence as the target endogenous gene. Sense gene constructs can be synthesized from mRNA obtained from a cell's cytoplasm using reverse transcriptase. Small changes can be made to the gene or multiple copies inserted into a vector for transfer. The effectiveness of these constructs varies depending on where in the genome they insert themselves, and the mechanism by which they prevent the endogenous gene functioning is still poorly understood.

An antisense gene has a coding sequence that is complementary to the target endogenous gene. This antisense gene can be manufactured on a DNA synthesizer and introduced in a vector. The antisense gene will be transcribed to give mRNA that is complementary to the mRNA transcribed from the endogenous gene. The mRNAs, being complementary, join or hybridize. The endogenous mRNA therefore becomes inactive and cannot be used at the ribosome to synthesise protein.

Gene-silencing technology was first commercially used in agriculture to create tomatoes with a higher solid content and a longer shelf-life, by preventing the synthesis of an enzyme involved in the ripening process (see Chapter 6). A range of slow-ripening fruit and vegetables is being developed using this technology. Gene-silencing technology has, however, a wide variety of applications. The main use of the technology is in medicine. Antisense DNA could be used to turn off critical protein synthesis in the development of cancers, AIDS, leukaemia and other harmful human gene activity.[19]

Plant tissue culture

Tissue culture is essentially a technique in which cells are grown on an artificial culture medium. These techniques are important in genetic engineering, both for the preparation of material to be exposed to foreign DNA and for the rapid production of whole organisms from transformed cells. Tissue culture techniques were initially developed during the 1950s, after it was observed that cells from both plants and animals could exist independently and be grown in glass flasks containing nutrients. Plant cells proved more versatile, however, as each plant cell has the potential for developing into an entire plant. Only the reproductive cells in animals have this potential.[21]

In plant tissue culture, a sterile sample of young actively growing tissue is usually taken, as this is less likely to contain bacterial, fungal or viral infection. The tissue is placed into a flask containing nutrient solution and plant hormones, chemicals that regulate the growth of plants.[11] A mass of undifferentiated tissue forms in the culture medium. Undifferentiated plant tissue can be transformed using bacterial vector or gene gun techniques. Samples of this tissue can be removed and placed on fresh medium, so that a large number of small clonal plants are produced. The process of producing whole plants from undifferentiated tissue is called *regeneration*.

Marker genes

As genetic engineering techniques developed, it became clear that only low success rates for transformation could be expected. In the case of transgenic crop plants, many unsuccessfully transformed plants need to be 'weeded' out to select useful transgenic plants. The low success rates are due to the relatively haphazard insertion methods currently available. Transferred DNA is inserted at random into the plant genome, while levels of transgene expression vary considerably between different independently transformed plants.[22] Many transformations are unstable and the interaction of the transferred gene with the plant genes is likely to vary depending on where in the genome it finds itself. Therefore, transformations require the incorporation of marker genes. These marker genes are transferred with the genes coding for the desired character and are closely linked to them in the recipient plant's genome.

An early marker was a gene expressing luciferase, obtained from the tails of fireflies (*Photinus pyralis*). This enzyme, in the presence of luciferin and a biochemical energy source, produces a light-emitting reaction. Fireflies use this light to signal to mates. Tobacco, transformed with the luciferase gene, glowed faintly when fed a luciferin substrate. Plants that had not successfully integrated the gene did not glow.[23] Luciferase genes can also be obtained from bacteria. This marker is one of a number of screenable or detectable markers, used to identify transformed plant material. Other screenable markers include beta-glucuronidase and beta-galactosidase, the expression of both being indicated by a blue colouration when tissue is incubated in the appropriate substrate for these enzymes.

Selectable marker genes not only allow transformed organisms or plant tissue to be identified, but also enable them to be selected from non-transformed organisms or tissue. Genes for resistance to a range of antibiotics became standard selectable marker genes during the late 1980s. The application of an antibiotic, usually under tissue culture conditions, kills non-transformed material. Antibiotic resistance marker genes are isolated from micro-organisms and express enzymes that degrade the corresponding antibiotic. Each antibiotic resistance marker gene is effective only against a limited number of antibiotics. The most commonly used selectable antibiotic marker gene expresses an enzyme called neomycin phosphotransferase (*npt* II; also known as *kan* or *neo*), which confers resistance to kanamycin, neomycin and other similar antibiotics. Other marker genes express enzymes that confer resistance to ampicillin and other penicillins (the *bla* gene), methotrexate (*dhfr*), hygromycin B (*aph* IV) and chloramphenicol (CAT).[24] Different crops have different natural resistance to antibiotics, for instance cereals to kanamycin, so a range of selectable marker genes have been developed for use in the production of transgenic crops.

Marker genes are also used to distinguish transformed from non-transformed bacteria, fungi, animals and fish. The same marker genes can be used in a wide range of organisms. Commonly used marker genes in fish, for example, include those expressing luciferase and neomycin phosphotransferase.[25]

Gene libraries

Plant collections storing seeds or cuttings are found in over sixty countries. These collections hold valuable crop genetic resources. For example, the International Rice Research Institute in the Philippines holds 60,000 rice (*Oryza sativa*) varieties. The economically important plants in these collections are now being used in research programmes, aimed at mapping and sequencing genes.

A gene map shows the relative positions of all the genes in an organism's genome. Until the 1990s gene mapping was a costly and time-consuming process. Gene-mapping techniques, originating in the 1920s, were based on observations that genes close together on a chromosome were linked and tended to be inherited together. Hence, linkage maps could be laboriously constructed to obtain the relative positions of genes. However, inexpensive and automated techniques for mapping genes are now widely available. Between 1920 and 1990, one hundred rice genes had been located. Between 1990 and 1994, ten to fifteen thousand rice genes were identified.

Gene libraries are collections of DNA fragments, representing the entire genome of an organism. They are created by breaking a genome into fragments using enzymes. These DNA fragments are then multiplied, by inserting each fragment into a single bacterium. Each subsequent bacterial colony contains identical copies of the original DNA fragment. Together these add up to a living gene library. These DNA fragments can be lined up to produce a map of the genome. Individual genes in a library can then be identified using a *gene probe*. A probe is a single-stranded piece of DNA, which is manufactured with a coding sequence complementary to that of the target gene. Probes can be labelled using radioactive phosphorus, which will darken photographic paper, or using fluorescent molecules, visible in ultraviolet light. All bands of DNA that bind to the probe can then be visualized.

Gene libraries contain the raw material resource for genetically engineering a particular crop, by facilitating the screening and isolation of genes. Taxonomists are also finding comparative gene sequences useful for constructing evolutionary lineages for a wide range of organisms. For example, the order of genes on the wheat and rice genomes are the same, reflecting their common cereal ancestor. One gene library, the GenBank database, already contained over 2,780 entries or sequences of plant origin by the early 1990s, including around two hundred distinct nuclear genes from higher plants.[26]

Gene sequencing involves the working out of the complete coding sequence of an organism's genome. The Human Genome Project, for example, aims to identify the complete human genome by the year 2005. The plant most frequently used as a research tool by molecular biologists is the Thale cress, *Arabidopsis thaliana*, which has a small genome and a rapid generation time.[27] It will be the first plant to have its genome fully sequenced. Techniques are being developed with Thale cress that will then be used to sequence the much larger genomes of the major crop species. The relative position of the genes for a number of cereals, such as barley, rye, wheat and millet, were already available by 1997. The complete gene sequences for several major food crops could be available by the turn of the century. Major projects to sequence the entire genomes of maize and rice have been initiated in the USA and Japan, respectively.[28]

Notes

1. Watson, 1968.
2. McClintock, 1951.
3. Ellis, 1983.
4. Vines, G., 1997, 'There is more to heredity than DNA', *New Scientist*, 19 April, p. 16.
5. Ho, 1996.
6. Tudge, 1993.
7. Pollack, 1994.
8. Rabinow, 1996. PCR was a key development in genetic engineering, made by Kary Mullis and colleagues at the Cetus Corporation in the USA in 1983. Nucleotides and DNA polymerase are put into a desktop machine and DNA amplification is an automated process. The polymerase is from the bacterium *Thermus aquaticus*, because the heat-stable enzyme from this species, which lives naturally in hot springs, survives the repeated heat-cooling process at the start of each cycle of DNA copying. The original story of the invention of PCR has Kary Mullis driving along a moonlit road in California when he was suddenly struck by a 'Eureka moment'. Rabinow's book, however, paints a more complicated picture involving team work and company politics.
9. Protein engineering will provide the foundation for the emerging engineering discipline of nanotechnology. For further information see: Drexler, K. E., 1992, *Nanosystems: Molecular Machinery, Manufacturing, and Computation*, John Wiley and Sons, New York.
10. White, 1993.
11. Grierson and Covey, 1988.
12. Brisson et al., 1984.
13. Fraley et al., 1983.
14. Jenes et al., 1993.

15. Potrykus, 1990.
16. Rhodes et al., 1988.
17. Klein et al., 1987.
18. McCabe et al., 1988.
19. Aldridge, 1996.
20. Grierson, 1996.
21. Research findings announced in early 1997 challenged this long-held view, when a sheep was cloned from an udder cell of an adult animal. Wilmut et al., 1997 (see Chapter 3).
22. Chyi et al., 1986; Dale et al., 1993.
23. Ow et al., 1986.
24. Kung and Wu, 1993a.
25. Iyengar et al., 1996.
26. Kung and Wu, 1993b.
27. *Arabidopsis thaliana* is a flowering plant in the family Cruciferae. It has only five pairs of chromosomes. Over half of this small genome codes for protein. It can be easily cultured and has a life-cycle of only six to eight weeks.
28. *Nature* 388: 312, 24 July 1997.

3. Milking it: increasing yields and the pharming of proteins

Milk is the essential first food of babies and forms a major component of children's diets. It is also an important part of the adult diet, through the consumption of cheese, yoghurt and other dairy products, made from the milk of cows and other farmed animals. Milk, the fluid secreted by the mammary glands of all mammals, provides a balanced and nutritious food for the young. Milk composition varies between species, with offspring being particularly adapted to the milk of their own species. Cows' milk is a more complex liquid than its popular image as a refreshing drink suggests. It comprises about 87 per cent water, 3.6 per cent lipids, 3.3 per cent protein (largely casein), 4.7 per cent lactose or milk sugar, and a range of vitamins (especially vitamin A and many B vitamins) and minerals, including calcium, phosphorus, sodium and potassium. Cows' milk naturally contains eight types of protein, three types of fat, eight minerals, fifty enzymes, sugars, eight vitamins and twenty-four hormones, including steroids and peptides.[1] Human milk contains less protein and more lactose than cows' milk.

Genetic engineering is now being used to manipulate both the quantity and quality of animal milk. Growth hormones produced by modified micro-organisms have been used to increase the milk yields of cows, whilst transgenic cows, goats and sheep have been produced that express additional proteins in their milk.

Recombinant bovine somatotropin (rBST)

Bovine somatotropin (BST), also called bovine growth hormone (BGH), is a hormone produced by the cow's pituitary gland, at the base of the brain, which is essential for growth, muscle development and milk production. A cow supplied with additional BST will produce substantially more milk. It was known as early as the 1930s that injecting cows with pituitary extracts could increase milk yields, and by the 1950s these effects were attributed to BST. However, until the arrival of genetic engineering and biotechnology the hormone could not be produced with sufficient purity or in large enough quantities for commercial use.

BST was to become one of the first genetically engineered biotechnology products for agriculture. Monsanto alone invested over US$1 million in its commercial production. The gene expressing BST was first isolated from cows and its coding sequence identified. Synthesized BST genes were then inserted into vectors and cloned in the bacterium *Escherichia coli*, using techniques similar to those developed for the production of insulin and other medicinal hormone products. The bacterial colonies were subsequently killed and the hormone extracted and purified.

Several companies developed slightly different recombinant bovine somatotropin (rBST) products. Monsanto's rBST was synthesized by the biotechnology company Genentech with a single extra amino acid; Dow Elanco synthesized its commercial BST with eight additional amino acids; American Cyanamid's product has three additional amino acids; and the Upjohn Company's product is identical to pituitary-derived BST. Naturally occurring BST has 191 amino acids. Minor changes to amino acid sequences arise due to the different manufacturing techniques employed and have value for patenting reasons if not for functional reasons. In all cases, the rBST is sold in sterile single-use syringes. Cows are injected once every 14 or 28 days and milk yields are increased by between 15 and 25 per cent.[1] Injections are timed to increase milk production during the later part of the lactation cycle.

Monsanto claimed that rBST was the most heavily researched veterinary medicine product in history.[1] The action of rBST in dairy cows has been studied in tests on over twenty-one thousand injected animals. The growth hormone is known to occur naturally in trace quantities in milk, and this amount was not increased using BST supplements. Monsanto concluded, from its data, that milk from treated and untreated cows was equivalent. BST is a protein, which is fully digested in the gut and is biologically inactive in humans, even if injected.[2] However, although the results from the trials appeared to give rBST a clean bill of health, doubts have persisted about the effects of long-term rBST use on animal health.

Selective breeding has enabled average milk yields per cow to increase, from around 1,000 litres in 1900 to 4,000 litres in 1990, pushing dairy cows towards their metabolic limit.[1] The use of rBST will push yields higher. This raises animal welfare concerns. Long-term rBST use is likely to lead to an increase in production-related diseases, including the serious udder infection called mastitis, and other metabolic and fertility disorders. Mastitis, which means 'inflammation of the udder', leads to a discoloration of the milk and can be detected by an increase in pus around the udder. Pus production is routinely quantified by a somatic cell count. The Ministry of Agriculture, Fisheries and Food (MAFF) in the UK until recently permitted somatic cell count levels equivalent to around 1 per cent of milk for human consumption.[3] Milk from cows with full-blown mastitis is, however, not permitted to be used for human consumption.

Monsanto's conclusion that rBST treatments did not increase the incidence of mastitis was contradicted by a re-analysis of its somatic cell count data by the independent researchers Erik Millstone and Eric Brunner.[4] Monsanto had analysed only the first 28-week period out of a 43-week study to arrive at its conclusion. This underestimated the apparent effect of rBST treatments, as these are most evident during the late period of lactation. The independent re-analysis of the data, for the entire period of the study, revealed that rBST treatments caused a 20 per cent increase in somatic cell counts compared to non-treated control animals, a highly significant difference. Monsanto re-analysed its data and published a further paper in 1994, which showed higher somatic cell counts than in the first published study.[4]

A number of hidden costs arising from long-term rBST use have often been ignored in reports and evaluations. The average life-span of cows is likely to decrease under a rBST treatment regime due to stress factors, for example, and cows will need to be replaced more frequently. A study conducted in Mexico, where Dow Elanco and Monsanto have aggressively marketed their rBST products at low prices, concluded that increased milk yields came at another cost, as cows eat additional dry matter to compensate for the higher milk production. The feed must also be an energy-dense mix, which is more expensive to buy. This study also concluded that injections had been administered more often than recommended, due to the dramatic increases in milk yields and competition-induced low prices.[5]

Although Monsanto concluded that no differences occurred in milk composition between treated and untreated cows, a number of published studies have shown an increase in fat concentration in milk from rBST-treated herds.[6] This coincides with a trend in the industrialized countries towards the consumption of low-fat milk, making any increase in fats undesirable. As milk production is increased due to rBST treatment, it is also possible that the proportion of certain vitamins and minerals will be lower.[6]

BST alters a cow's nutrient balance in favour of milk production, with subsequent changes in other tissues. These changes in tissue metabolism are mediated by another group of hormones called insulin-like growth factors (IGFs). These hormones are naturally present in milk, although one of this group of proteins, insulin-like growth factor I (IGF I), is present at higher levels in milk from rBST-treated cows than in milk from untreated cows.[7] IGF I is denatured during the processing of cows' milk for infant formula.[7] IGFs are naturally present in the human body, and human and bovine IGF I are chemically identical. However, IGF I stimulates cell division and, in theory, excessive levels could promote cancer. Statements in an official MAFF report in 1994 led a molecular biologist at Cambridge University, Paul Schofield, to suspect that IGF I may be a greater matter for concern than previously thought.[4] The MAFF report stated that IGF I is digested in the gut, that no receptors for IGF I existed in the human gut, and therefore IGF I in milk is rendered safe as far as human health risks are concerned. Published research, however, existed showing that both statements, on which MAFF based its conclusions, were untrue. A protein has been found that appears to protect IGF I in the human gut, keeping it active, while receptors for the hormone have been identified.[4] The higher levels of IGF I in milk from rBST-treated cows are, however, still within the normal physiological range of human breast milk.[7] They may not, therefore, present a significant health risk to consumers.

In the USA, the Food and Drug Administration (FDA) concluded that cows' milk from rBST-treated herds was safe to drink, basing its findings on extensive data collected from four companies. These data included a study that concluded that additional IGFs would not be absorbed by humans drinking this milk.[7] In 1986, the FDA allowed Monsanto, Dow Elanco, American Cyanamid and Upjohn to sell the milk, and cheese made from it, from their rBST trials. This enabled the companies to recoup some of the costs of rBST development. The milk was not labelled as being from rBST-treated cows, even though rBST was at that time an unlicensed drug. It is not known who drank the milk as secrecy prevented this information being revealed. In November 1993, the FDA approved Monsanto's rBST for the US marketplace. The Monsanto product is marketed under the Posilac brand-name in the US and elsewhere.

Although it was welcomed by the many in the US dairy industry, rBST was boycotted by a number of dairy farmers, four major supermarket chains and a large number of consumers.[6] The dairy farmers were concerned that increased milk production might decrease prices, forcing smaller operators out of business. Higher yields would also mean that fewer cows would be needed to produce the same amount of milk, putting further pressure on smaller producers. If rBST were to be widely adopted in the USA, it is estimated that the number of cows needed to meet the country's milk requirements would drop from 10.8 million to 7.5 million, and the number of dairies would be halved.

By the mid-1990s, the FDA acknowledged that rBST was causing unexpected problems. The main reason for the FDA's concern was that cows ate large amounts of food under rBST-treatment regimes, causing stress on the immune system and leading to more infections, including infections of the udder such as mastitis. Cows with udder infections need more antibiotics, which can get into the milk. Milk and milk products from rBST-treated animals are therefore more likely to contain residual antibiotics. These antibiotics can get into the human gut with the milk and may select for antibiotic-resistant bacteria. Micro-organisms responsible for illness, for example certain strains of *E. coli*, might then become more resistant to the antibiotics used to treat them.

In the UK, MAFF also concluded, from reviewing trial data, that milk from rBST-treated cows was safe to drink.[6] Periodic MAFF reviews during the late 1980s, and again in 1993, resulted in the same conclusion. However, its Veterinary Products Committee (VPC) in 1990 rejected, on animal health grounds, a Monsanto application concerning marketing of rBST, despite considerable political pressure for it to be approved.[4] Monsanto re-applied to another regulatory committee, the Medicines Committee, which overturned the VPC decision in 1993. The Medicines Committee also gave marketing approval to Dow Elanco's rBST. These committees operated under conditions of secrecy, so no details of the decision-making process were made public.[8] The data that companies submit for marketing approval are not available for public scrutiny.

The MAFF decision led to milk from cows used in experimental trials of rBST being added to the general milk supplies by the Milk Marketing Board during 1987 and 1988. The milk from the trials was treated like milk from other sources, and was not labelled as being from rBST-treated cows. The dairy industry responded to critics by pointing out that regulations around the world backed the view of MAFF that the milk was safe, so, by extension, a challenge to the UK regulations was a challenge to regulations worldwide. However, the pooling of milk from rBST-treated herds was not done in any other European country.[6] The high level of secrecy within MAFF and the dairy industry has prevented the full scale of the supplies of milk obtained, or the locations of the herds used in the rBST trials, being known. However, it was estimated that around three thousand cows were involved in the trials, which were probably conducted at the government's Shinfield and Hurley research institutes, at London

University's Wye College, and on commercial dairy farms in Devon, Somerset, Dorset, West Wales and Yorkshire.[4] The experimental trials were designed to study the long-term effects of rBST injections on cow health. No studies were planned on the long-term effects on humans of consuming milk from rBST-treated cows.

Monsanto, American Cyanamid and Dow Elanco applied for marketing approval for their rBST products in the European Union (EU) in the late 1980s. Monsanto presented its first application in July 1987, but a decision was delayed until 1991 because several member states had reservations about its use. An initially favourable response was obtained, but a moratorium on its use in Europe until the year 1999 was subsequently imposed. The decision was influenced by the fact that milk is overproduced in Europe, quotas having been in effect since 1984. In addition, people in Europe showed no indication of increasing their milk consumption; in Britain, for example, during the period 1984 to 1994 milk consumption dropped by 10 per cent, with a shift towards lower-fat milk.[9] Therefore, increased milk yields were not seen as a desirable objective.[10]

When they learned that milk from rBST trials was being sold for human consumption the Retail Consortium in the UK, which includes Co-op Dairies, Express Dairies, Marks and Spencer, Sainsbury, Tesco and Waitrose, wrote to MAFF in 1988 asking for the consumer's right to choose to be respected.[1] This would have required separate farm collections and labelling of milk from rBST-treated cows. In addition to economic factors, the dairy industry feared that labelling would be bad for sales. The request was rejected, and milk from rBST-treated herds may again be unlabelled after the moratorium on its use in Europe ends in 1999, despite stricter requirements for labelling being imposed in 1997 (see Chapter 13), as it may fall outside the definition of a genetically modified food. In addition, Monsanto has asked, through the US government, that the World Trade Organization (WTO) rule the European ban on rBST illegal (see Chapter 10). rBST products are likely to be on the European market by 1999, although they suffered a setback in June 1997 when the Codex Commission, the international food standards body,[11] failed to pass a vote approving the use of rBST in cows.[12] Although these standards are not binding, they are used for international trade purposes by many countries. However, in August 1997 the WTO ruled that the EU could no longer exclude meat and milk from cattle treated with BST.[12]

Recombinant BST products have been aggressively marketed in the developing world. Eli Lilley and Monsanto, in collaboration, have particularly targeted the Indian market. India is the second largest producer of milk in the world, with one-third of the world's dairy cattle, but the yield per cow is the smallest in the world.[12] Multinationals are also actively marketing rBST in Central and South America.

The growth hormones of other domesticated animals have also been identified and integrated into bacteria for commercial production. Monsanto, for example, may soon seek marketing approval for recombinant porcine somatotropin (rPST), for injection into pigs to produce leaner pork.

The pharming of transgenic animals

One of the aims of transgenic animal research has been the production of additional proteins in the milk of mammals, particularly proteins that can be used as pharmaceutical drugs. This has involved integrating genes expressing human proteins, into the embryos of other mammals. A female mammal can yield far greater amounts of protein in her milk than could be obtained from genetically manipulated bacterial cells grown in a fermenter. Goats, cows and sheep have been modified as highly efficient living pharmaceutical factories, continuously producing drugs in their milk, within a new industry called 'Pharming'. Some of the products of pharming may be sold as 'neutraceuticals': part food and part drug. Milk from transgenic animals could, for instance, have enhanced vitamin content or contain other dietary supplements.

Transgenic animals are produced by micro-injection, with the foreign genes being injected directly into a fertilized egg using a miniature pipette. It takes many attempts to produce a transgenic animal. The micro-injection method is inefficient, with fewer than 20 per cent of eggs taking up foreign genetic material. Of animals born live, only about 10 per cent will be transgenic and of these only 1 per cent are likely to express the transgene to the appropriate level.[13] In another estimate, it was calculated that for every 10,000 eggs injected with foreign DNA, only about three are likely to reach adulthood and express the required protein in sufficiently large quantities. These animals, with a functional transgene, are called 'founder animals' and are extremely valuable. The United States Department of Agriculture (USDA) has estimated that the costs of a single founder pig, sheep and cow are US$25,000, US$60,000 and US$300,000 to $500,000, respectively.[13] They are treated like royalty, given cute names and often become media stars. Rabbits have been used in some of the initial research to reduce costs, and it has been suggested that their milk could be of commercial value in some cases. Methods of identifying transgenic embryos are continually improving, for example by using fluorescent markers, such as a gene from jellyfish which gives a green glow, and costs of producing transgenic animals are likely to decrease in the future.[13]

One of the first transgenic goats out of the laboratory was called Grace. In her first year she produced about one kilogram of a chemotherapeutic drug.[14] Such animals have great commercial potential. The biotechnology company PPL Therapeutics was set up in 1987, close to the Roslin Institute in Edinburgh, Scotland, to commercialize the research of the institute, which has made a series of breakthroughs in the area of animal genetics.[15] A cow named Rosie, from the Roslin Institute and PPL Therapeutics stable, expressed in her milk alpha-lactalbumin, a protein found in human breast milk. This enriched cows' milk may be sold for its increased nutritional value, but its most useful application will be as a formula product for premature babies.[16] The infant formula market is worth US$4 billion annually worldwide. In 1990, a sheep called Tracey, also from the Roslin Institute, produced abundant supplies of a drug called alpha-1-antitrypsin.[17] A deficiency of this compound in humans leads to the lung disease emphysema, which affects about one hundred thousand people in the western world. It is estimated that a herd of a thousand sheep would produce enough of this protein in their milk to satisfy world demand.[18]

GenPharm International, a multinational with offices in the USA and The Netherlands, also has cows expressing human milk proteins in their milk. Their first transgenic success was Herman, a bull born in 1990, whose female offspring produce human lactoferrin in their milk.[5] The process by which transgenic cows express two human proteins, lactoferrin and lysozyme, not currently present in infant formula milk, has been patented. These proteins have specific iron-transport and anti-bacterial properties, respectively. GenPharm has a collaborative agreement with Bristol-Myers Squibb to market a nutritionally enriched infant formula worldwide.

Dolly the sheep and animal cloning

Pharmaceutical companies plan to be milking herds of transgenic goats, cows and sheep for their valuable protein products in the near future. However, the production of transgenic animals is a hit-and-miss affair, and flocks bred from founder animals are variable, including high and low producers of the desired proteins. The low success rate can prove costly, and has driven researchers to further pioneering work. The production of transgenic herds of cloned animals came a step closer when it was announced, in February 1997, that a sheep called Dolly had been cloned from a cell taken from the udder of a six-year-old ewe.[17] This research, carried out at the Roslin Institute in collaboration with PPL Therapeutics, was of great significance to the study of genetics and animal development. The result contradicted conventional thinking on cell differentiation in animals. It had been thought that all animal cells, except the reproductive or germ cells, were irreversibly differentiated from the embryo stage onward. The udder cell that produced Dolly, however, was manipulated so that it returned to a primitive or undifferentiated state, from which a whole organism could be regenerated.

The animal cloning research has major financial implications, as the market for therapeutic proteins, which in 1997 was about US$7.6 billion a year, is expected to rise to US$18.5 billion by the year 2000.[19] Breeding founder animals with other animals using traditional animal-breeding techniques may dilute an introduced transgene's effect and is a slow process. However, now founder animals can be cloned to produce herds of identical animals expressing a standard drug product. Dolly was born in July 1996, six months before her birth was publicly announced. The announcement of her birth was delayed because patent applications had to be filed before any results could be published. These patents are of a broad nature and cover the use of the techniques with all mammals, including, by extension, humans. PPL Therapeutics has agreements with at least four major pharmaceutical companies – Novo Nordisk, American Home Products, Bayer and Boehringer Ingleheim – to market therapeutic drug products from transgenic and cloned animals.[19] In July 1997, the Roslin Institute announced the birth of a lamb named Polly, the first cloned animal to contain a human gene expressing a protein for a therapeutic drug.[20]

Cows have also been cloned, from the cells of embryos. A company called ABS Global, in the USA, announced a new technique for mass-producing clones in August 1997, after they had introduced a cloned calf called Gene to the media.[12] PPL Therapeutics is seeking to introduce genes expressing human proteins into cloned cows. A company called Pharming, based in Leiden, The Netherlands, is also trying to produce pharmaceutical drugs in cows with similar techniques.

The cloning of Dolly proved controversial, however, with calls for bans on cloning experiments from various quarters. The media, at the time, made much of the potential for human cloning, and little of the immense medical benefits that such work could lead to. Laws banning human cloning experiments were rushed through in the USA and other countries. The geneticist Steve Jones, commenting on the media's response to the cloning experiments, said that 'the public is not frightened of progress but rapid progress'.[21] The cloning of Dolly has amazed many scientists, so it is not surprising if the general public are slow to grasp the implications of this research. Regulations may help put a brake on this work, to slow it to a speed acceptable to the public. There are immense benefits to be gained, as well as immense profits, but there should be open public debate and close monitoring of how the technology is being used. In this controversial and sensitive area of science, it would be wrong to proceed with commercial applications, using the cloning and genetic manipulation of mammals, if the majority of people are unwilling to support such developments.

Notes

1. Deakin, 1990.
2. Miller, H., 'Putting the bSt human-health controversy to rest', *BioTechnology*, 10: 147, February 1992.
3. MAFF permitted a somatic cell count of 400,000 cells per millimetre of milk. Monbiot, G., 'Agribusiness uncowed by suffering', *Guardian*, 9 July 1997, p. 17.
4. 'How safe is safe?' *Sci Files*, BBC2, transmission 14 April 1997.
5. RAFI, 1993.
6. Brunner, 1990.
7. Juskevich and Guyer, 1990.
8. The VPC advises MAFF on the licensing of new veterinary drugs. The committee has around twenty members, who meet eleven times a year to assess around one hundred products. It operates under Section 118 of the 1968 Medicine Act, under which committee members are under threat of two years' imprisonment if they reveal details of the decision-making process.
9. *Britain 1996: An Official Handbook*, HMSO Books, London, 1995.
10. Vandaele, W., 'bST and the EEC: politics vs. science', *BioTechnology*, 10: 701, February 1992.
11. The *Codex Alimentarius* Commission was set up in 1962 by the Food and Agriculture Organization (FAO) and the World Health Organization (WHO). See also Chapter 13.
12. *Nature Biotechnology*, 15: 701, August 1997; *Guardian*, 17 September 1997.
13. *Nature Biotechnology*, 15: 416, May 1997.
14. *AgBiotech: News and Information* 8 (9): 155N, September 1996.
15. PPL Therapeutics also has a research station in Blacksburg, Virginia, USA, where micro-injections are carried out, and where Rosie was born.
16. *New Scientist*, 15 February 1997, p. 12.
17. Wilmut et al., 1997.
18. Aldridge, 1996.
19. RAFI, 1997.
20. *New Scientist*, 2 August 1997, p. 17. Polly was produced from an embryo modified with a human gene, and then cloned by inserting genetic material into sheep's eggs that had their DNA removed. This resulted in the birth of transgenic cloned lambs. Polly had two clone 'sisters'. Most transgenic animals produced by the Roslin Institute in the future are likely to be produced by this type of technique, rather than cloning from adult animal cells.
21. *Nature* 386: 8–9, 6 March 1997.

4. Herbicide-resistant crops

Herbicide resistance is the characteristic most commonly engineered into transgenic crop varieties grown in field trials. By 1987, over twenty-eight companies had already launched research programmes in herbicide resistance.[1] The seven leading agrochemical producers, accounting for over 60 per cent of the world market, are developing herbicide-resistant crops.[2] These crops represent the most profitable use of genetic engineering in crop production to date, because herbicide-resistant crops generate demand for herbicides. Sales of herbicides worldwide amount to almost US$5 billion annually, which is about 40 per cent of total pesticide sales.[3] In addition, seed companies are increasingly being acquired by the agrochemical-producing multinational companies. The sale of genetically modified seeds and increased herbicide sales are estimated to be worth at least US$6 billion by the year 2000.[1]

Advantages for weed control

Weeds compete with crops for moisture, nutrients and light, and therefore uncontrolled weed growth can result in large yield losses. The presence of weeds at harvest can also decrease crop quality, for example by decreasing the purity of grain. Herbicides have played a major role in increasing crop yields since the Second World War, although monetary losses due to weeds can still be as high as 10 to 20 per cent of crop value.[4] Broad-spectrum herbicides are effective against a wide range of weed species, but can kill or injure crops at the levels required for effective weed control. Herbicide applications are constrained by the damage they do to the crops themselves, and herbicides could be used much more efficiently if crops were made resistant to them. Conventional plant breeding has addressed this problem, but with limited success. Studies of naturally occurring herbicide resistance in the field have shown that resistance is usually due to a single mutation and therefore presents a good target for genetic manipulation.

Besides damaging crops onto which they are directly sprayed, herbicides can also damage crops planted in soil that has received herbicide sprays earlier in a crop rotation. Atrazine, for example, is a widely used herbicide on maize, a crop that has a natural tolerance to it. However, this herbicide is persistent and may remain active in the soil for long periods. Soybean, often grown in the soil after maize, is very sensitive to atrazine. The development of atrazine-resistant soybeans would, therefore, enable more applications of atrazine to be made to preceding crops without adversely affecting soybean. It would also allow atrazine to be used to give greater levels of weed control in soybeans.[3] Therefore, herbicide-resistant crops allow greater flexibility in the choice of crops and herbicide treatments during a rotation.

Achieving herbicide resistance

Most groups of herbicide are broken down naturally in the field by soil bacteria. This has been exploited by genetic engineers, who transfer genes for detoxifying enzymes from soil bacteria into transgenic crops. Genes for these enzymes can also be isolated from plants that are naturally resistant to particular herbicides. Most herbicide-resistant crops were initially developed using *Agrobacterium* to integrate foreign genes into plant cells. However, particle bombardment methods are now routinely used, particularly with cereal crops and soybeans (see Chapter 2).

The integrated foreign genes, from bacteria or other plants, can do one of several things to achieve herbicide resistance. They can, for example, express proteins that degrade or detoxify the herbicide, or they can alter the sensitivity or quantity of the enzymes that the herbicide acts upon to kill the plant.[4] Resistance obtained by genetic modification is usually to a single group of herbicides. Plants engineered for resistance to one herbicide will not necessarily be resistant to other herbicides, because different herbicides have different modes of action. Bromoxynil and atrazine are inhibitors of stages in the biochemical pathways of photosynthesis, 2,4-D[16] is a plant growth regulator, and other herbicides inhibit various steps in amino acid biosynthesis.

Resistance has been obtained to most of the major herbicide groups, although research has concentrated on certain herbicides during the development of transgenic crops. In a survey of field releases of transgenic plants in the industrialized countries from 1986 to 1992, herbicides to which resistance had been engineered included glyphosate (190 releases), glufosinate ammonium (188), sulfonylurea (64) and bromoxynil (27).[5]

Glyphosate[6] is an organophosphorus compound, i.e. a synthesized organic chemical containing phosphorus, which acts as a broad-spectrum, non-selective, post-emergent herbicide.[7] It can therefore be used to control most of the major weed species found in crops. Plants sprayed with glyphosate rapidly transport the herbicide to their growing points, where it acts by inhibiting an enzyme called EPSPS.[8] This blocks amino acid biosynthesis and leads to cessation of growth and eventual plant death.

Glyphosate tolerance can be engineered into crops using genes from bacteria or plants. The first transgenic glyphosate-resistant plant was tobacco, incorporating a gene from the bacterium *Salmonella typhimurium*. This gene expressed a version of EPSPS that was not susceptible to glyphosate.[9] These plants were herbicide-resistant in the field, but did not grow as well as untreated controls.[10] Another approach was to introduce genes from a variety of *Petunia hybrida* that had been artificially selected for glyphosate tolerance. This plant variety overproduced EPSPS, owing to a twenty-fold amplification in the number of copies of the gene. A vector construct containing these multiple gene copies was transferred into other *Petunia* plants. These plants withstood four times the amount of glyphosate required to kill a non-modified Petunia of the same type.[11] Monsanto produces the glyphosate herbicide Roundup™. The production of crops resistant to Roundup™ will be discussed later in the chapter.

Glufosinate ammonium is in the phosphinothricin (PPT) group of herbicides.[12] Hoechst's Basta™, a widely used glufosinate ammonium herbicide, was introduced to the market in 1981 for use against both narrow- and broad-leaved weed species. Hoechst's crop production division merged with that of Schering's in 1994 to form AgrEvo. All PPT herbicides inhibit an enzyme (glutamine synthase) involved in the assimilation of ammonia, which is used by plants in the synthesis of the amino acid glutamine. This enzyme plays a key role in regulating nitrogen metabolism in plants, and its inhibition results in ammonia building up to toxic levels.[13]

Genes from alfalfa (*Medicago sativa*) and soil bacteria have been used to produce transgenic crops resistant to Basta™ and other glufosinate ammonium herbicides. A mutant gene from alfalfa, expressing glutamine synthase, has been used to obtain some resistance in transgenic tobacco. This approach relies on an overproduction of enzyme to counter the herbicide's enzyme-inhibiting effect. A more promising approach, however, was using a gene, called the *bar* gene, from the bacterium *Streptomyces hygroscopius*, which expresses an enzyme (PPT acetyltransferase) that effectively detoxifies the herbicide by altering its chemical structure.[14] Plant Genetic Systems, in collaboration with AgrEvo, has produced a number of different crops with resistance to glufosinate ammonium herbicides. Ciba-Geigy's insect-resistant maize also includes a gene conferring resistance to Basta™ (see Chapter 5).

Two related herbicide groups, active against broad-leaved weeds in wheat, rice, soybean and other crops, are the sulfonylureas and imidazolinones. They have a broad-spectrum action, but are effective at low spray rates and have a relatively low toxicity to animals. The sulfonylureas, developed by Du Pont, were introduced to the market around 1980. For example, Glean™ is a sulfonylurea herbicide used in wheat, a crop with a natural resistance to it. Sulfonylureas are toxic to weeds because they inhibit an enzyme (acetolactate synthase) involved in the biosynthesis of amino acids (leucine, valine and isoleucine).[13] Resistance to sulfonylurea herbicides can be obtained by transferring genes for this enzyme from plants that produce it in abundant quantities, for example, *Arabidopsis thaliana.*[15] This causes an overproduction of the enzyme in the transgenic plants, neutralizing the herbicide's toxic effect.

The imidazolinones, developed by American Cyanamid, also inhibit the same stage of amino acid biosynthesis. The same genes can, therefore, be transferred to obtain resistance to both these herbicide groups. Imidazolinones are active against both broad- and narrow-leaved weeds, but are selectively toxic due to differential rates of herbicide metabolism between crop and weeds, which makes them particularly useful in cereals for the control of grass weeds. They are used in soybeans, but their persistence in the soil can harm crops grown after soybeans in crop rotations. American Cyanamid and Du Pont are therefore developing resistance in a range of other crops, to enable them to thrive in rotations with soybean. American Cyanamid has licensed a resistance gene to Pioneer Hi-Bred for incorporation into their transgenic varieties of maize, which will confer resistance to American Cyanamid's imidazolinone herbicides.

Bromoxynil, one of a group of herbicides referred to as nitriles, is the active ingredient in Buctril™, a herbicide produced by the French-based multinational Rhône-Poulenc. This herbicide is used for broad-leaf weed control in maize and wheat, both crops with some natural resistance to it. Resistance to bromoxynil is obtained by transferring a gene, isolated from a strain of the soil bacterium *Klebsiella ozaenae*, which expresses a bromoxynil-specific nitrilase enzyme in crop plants.[11] This enzyme converts bromoxynil to an inactive chemical. Calgene have produced transgenic cotton resistant to bromoxynil using their patented BXN™ gene. More herbicides are sprayed on cotton (*Gossypium* spp.) than probably any other crop, but weed control in cotton is constrained by the lack of a broad-leaf post-emergent herbicide that will leave cotton undamaged. Bromoxynil-resistant cotton fulfilled this need, when it was commercially introduced in April 1995. In 1996, BXN™ cotton was grown on 20,200 hectares, and in 1997 on around 178,000 hectares. Calgene sells this BXN™ cotton seed at a premium price: 41 per cent more than their non-modified seed.

Triazine herbicides, produced by Du Pont and Ciba-Geigy, act to disrupt photosynthesis by interfering with a process involving the binding of proteins in the chloroplast. Triazine-resistant plant species have arisen spontaneously, with a differently structured protein whose binding is unaffected by these herbicides. Genes from these species can be used to produce herbicide-resistant crop plants. Resistance to the triazine herbicide atrazine, for example, has been obtained using a mutant gene from *Amaranthus hybridus*.[13] Du Pont has produced atrazine-resistant soybeans, which could increase atrazine sales by US$120 million annually.[3] However, a major drawback to the production of plants resistant to triazine herbicides is that the gene encoding the protein is situated on the chloroplast DNA. Foreign genes are usually integrated into the DNA in a cell's nucleus, and it is more difficult to introduce them into chloroplast DNA. The mutant gene from *Amaranthus* had to be converted into a nuclear gene, using different regulator genes, to be effective.

The triazines are persistent herbicides, which is advantageous in that they could provide effective weed control throughout a rotation of different triazine-resistant transgenic crops, but could be disadvantageous in that these herbicides may be highly damaging to the environment if used in this way. Although much research was done on obtaining resistance to them during the 1980s,[3] few experimental releases of triazine-resistant crops have been made in recent years.[5]

Plant growth regulators can also be used as herbicides. 2,4-D occurs naturally as a plant hormone that stimulates cell growth. However, in large quantities this chemical kills plants by promoting excessive growth. It was developed as a herbicide by Schering Agrochemicals and is used on cereals, as cereal crops can metabolize it while broad-leaved weeds cannot. Resistance to 2,4-D has been obtained using a gene, isolated from the soil bacteria *Alcaligenes eutrophus*, which expresses an enzyme called DPAM[17] responsible for degrading 2,4-D to an inactive chemical.[11] AgrEvo are developing 2,4-D-resistant maize.

Monsanto's Roundup Ready™ crops

The world's biggest-selling herbicide is Monsanto's Roundup™, whose active ingredient is glyphosate. As previously mentioned, glyphosate acts by inhibiting an enzyme called EPSPS,[8] leading to disruption of amino acid biosynthesis. To develop their transgenic Roundup Ready™

soybeans, Monsanto used mutant genes from bacterial strains of *Pseudomonas* spp. and *Klebsiella pneumoniae*.[4] These mutant genes expressed EPSPS, leading to an overproduction of the enzyme in the plant, which counteracts the suppression of this enzyme by glyphosate. Monsanto has subsequently incorporated these patented Roundup-tolerant genes into a range of other crops, including maize, canola (spring rape), oilseed rape, sugar beet, tobacco and cotton.

The first field trials with herbicide-resistant soybeans were conducted in 1989 and 1990. These transgenic plants showed significant tolerance, but it was only after 1991 that transgenic soybeans demonstrated commercially useful levels of herbicide resistance with no loss of yield.[18] The first major commercial planting of Roundup Ready™ soybean was in 1996, when transgenic seed accounted for around 2 per cent of the total US crop. The proportion of the crop produced using transgenic seed increased to around 15 per cent in 1997 and is set to rise further in the coming years.

Monsanto scientists have published data showing that the composition of glyphosate-resistant soybean seeds is equivalent to that of conventional soybeans.[19] Their data suggested that foods produced using modified soybeans should be no different from foods produced using unmodified soybeans. An important use of soya in the US is in animal feed. The feeding value to animals was shown to be unaffected by the incorporation of the glyphosate-resistance gene,[20] while the protein expressed by the foreign gene was rapidly digested by mice.[21]

Monsanto has released herbicide-resistant crops, under experimental conditions, around the world. In Britain, for example, Monsanto has grown Roundup Ready™ sugar beet since 1995 under field trial conditions in sites throughout south-east England.[22] The transgenic sugar beet was able to withstand three times the normal rate of Roundup sprays, without damage to the crop, with yield increases of up to 7 per cent. A Monsanto spokesman claimed it would have taken plant breeders, using traditional methods, twenty years to achieve a similar result. Trials in Britain with Roundup Ready™ oilseed rape also started in the mid-1990s.[23]

Monsanto's Roundup Ready™ soybeans were among the first genetically modified organisms to be widely marketed as ingredients for a range of foods (see Chapter 12).

Environmental considerations

Theoretically, resistance to any herbicide could be engineered, but in practice various factors are taken into account. Glyphosate, for example, according to Monsanto has several desirable properties for a herbicide used on herbicide-resistant crops, including a broad-spectrum action, high unit activity, low volatility and soil mobility, and relatively low toxicity to aquatic life, birds and mammals. The potential for weeds to acquire tolerance to glyphosate by gene spread should also be low.[4]

However, herbicide-resistant crops could themselves become weeds in other crops, while related weedy species could acquire resistance through pollen transfer from transgenic crops. Certain crops, and certain types of herbicide resistance, present greater ecological risks. A project to obtain sulfonylurea resistance in oilseed rape, for example, was stopped when it was realized that volunteer rape would itself become a weed in wheat, where it would be resistant to the most important herbicide used in that crop.[13] Transgenic oats or sorghum could interbreed with wild oats or johnson grass, potentially spreading herbicide resistance to weed species. Each case for a transgenic herbicide-resistant variety therefore needs to be assessed individually for risks of increased invasiveness or possible transgene spread (see Chapter 7).

Herbicide-resistant crops are likely to increase the amount of herbicide sprayed into the environment. Monsanto claims, however, that the use of herbicide-resistant crops will decrease the number of herbicide sprays required and will promote environmentally sound herbicide usage.[4] Monsanto argues that by using a herbicide-resistant crop, a single herbicide spray could be used to kill all weeds after the crop has started to emerge, including types of immature weeds that would normally have required spraying just before crop emergence. The agronomic utility of broad-spectrum herbicides is, therefore, likely to be increased. It should also be noted that some conventionally bred varieties have been produced that are resistant to herbicides, so the argument that the utilization of resistant varieties will lead to increased herbicide use is not just an argument against genetically modified varieties.[24]

Herbicide-resistant crops may lead to a more effective use of herbicides, but the argument that they will not lead to increased use of herbicide is difficult to sustain. Indeed, from a commercial point of view, the aim has always been to sell more herbicide. Application of herbicide can now be done in certain crops, with engineered resistance, and under certain circumstances, when spraying was previously not possible. Multinational companies have been applying to have the range of uses of major herbicides officially extended to cover these new opportunities. Previously, an upper limit existed on the herbicide spray rate, because above that level crop damage occurred. With herbicide-resistant crops, however, a tendency may exist to overspray as there is unlikely to be an adverse effect on the crop plants. Any increase in herbicide spraying may lead to increased herbicide residues in food. Monsanto has made applications to the Australian and New Zealand governments to increase the permitted levels of Roundup herbicide residues in soybeans, following the importation of Roundup Ready™ soya into these countries.[25] Meanwhile, increased amounts of glyphosate herbicide, being sprayed on Roundup Ready™ cotton in the USA, may be getting into cotton seeds, the oil from which is used in a range of food products.[25] The planting of herbicide-resistant transgenic crops is likely, therefore, to lead to increased amounts of potentially hazardous herbicide being sprayed onto crops.

Increased use of herbicides, due to widespread deployment of herbicide-resistant crops, could have a number of undesirable environmental effects. Herbicides can have adverse ecological effects on natural habitats near farmland. Glyphosate is a non-selective herbicide, for example, which is lethal to a wide range of herbaceous plants. The US Fish and Wildlife Service identified 74 endangered plant species potentially threatened by excessive glyphosate use. Adverse effects on soil fertility may also occur. For instance, glyphosate is suspected of inhibiting the growth in the soil of mycorhizal fungi, which help plant roots absorb minerals from the soil. Therefore, glyphosate herbicides, including Roundup, are not environmentally friendly chemicals. Increased glyphosate use might also have direct adverse effects on human health. In a Californian study, glyphosate was identified as the third most common cause of pesticide poisoning among farm workers.[26]

An overuse of herbicides may prove counterproductive for other reasons. For example, aphids became more numerous on maize sprayed with 2,4-D, probably because of changes in the plant's sap.[27] Increased herbicide use may therefore lead to increased use of insecticide sprays in such cases. Du Pont has conducted research into plant tolerance to the herbicide picloram, but found that sugar levels in roots increased, creating favourable conditions for the growth of disease-causing bacteria and fungi.[3] Increased rates of herbicide spraying also select for herbicide-resistant weeds. The development of unwanted herbicide resistance in common weed species is a growing problem in agriculture. For example, blackgrass has developed resistance to herbicides used in cereals. Any increased herbicide usage may therefore increase the rate of development and spread of herbicide-resistant weeds, cancelling out the initial benefits of transgenic crops.

There is immense potential for herbicide-resistant crops to improve weed management and crop yields, while providing a more cost-effective and arguably more environmentally acceptable weed control. A desirable future objective is to produce herbicide resistance in crops to control parasitic weeds, such as dodder (*Cuscuta* spp.), broomrape (*Orobanche* spp.) and witchweeds (*Striga* spp.). At present, no herbicide has a sufficient margin of selectivity to deal with these weeds without damaging crops.[28] Herbicide resistance genes can also be useful in transgenic plants as selectable markers, in combination or instead of antibiotic resistance marker genes. These markers provide a screen or assay for potentially transformed plants, which alone will survive a herbicide treatment. A herbicide resistance gene, for example, was used in the production of Ciba-Geigy's *B.t.* maize. Herbicide-resistant crops represent a high-input solution to weed control, however, which is not compatible with current ideas of sustainable agriculture. Transgenic crops will need to be used carefully if problems of weed resistance to herbicides and ecological damage are to be avoided.

Notes

1. Juma, 1989.
2. Bayer, Ciba-Geigy (Novartis), Du Pont, Hoechst (AgrEvo), Monsanto, Rhône-Poulenc, Zeneca.
3. Hobbelink, 1991.
4. Hinchee et al., 1993.
5. Landsmann, Shah and Casper, 1995.
6. Glyphosate may be referred to by its chemical name N-phosphonomethyl glycine.
7. Grossmann and Atkinson, 1985.
8. 5-enolpyruvylshikimate-3-phosphate synthase. This enzyme interrupts the metabolic pathway that incorporates shikimic acid for the production of aromatic amino acids.
9. Comai et al., 1985.
10. Quinn, 1990.
11. Dale et al., 1993.
12. PPT herbicides are chemically synthesized analogues of L-glutamic acid.
13. Oxtoby and Hughes, 1990.
14. De Block et al., 1987.
15. Miki et al., 1990.
16. 2,4-dichlorophenoxyacetic acid.
17. 2,4-dichlorophenoxyacetate monooxygenase. This enzyme converts 2,4-D to 2,4-dichlorophenol.
18. Delannay et al., 1995.
19. Padgette et al., 1996.
20. Hammond et al., 1996.
21. Harrison et al., 1996.
22. For example: Department Of the Environment (UK), Public Register of Genetically Modified Organism Releases: 96/R22/5, 96/R22/4, 95/R22/2. http://www.greenpeace.org.uk/science/ge/releases.html
23. *Farmers Weekly*, 7 March 1997, p. 62.

24. Caseley et al., 1991.

25. Greenpeace Press Release, Auckland, New Zealand, 20 April 1997; *Nature Biotechnology* 15: 1233, November 1997.

26. *AgBiotech: News and Information* 8 (12): 197N, December 1996.

27. Oka and Pimentel, 1976.

28. Gressel, 1993.

5. Insect-resistant crops and a modified insect baculovirus

Many of the early experiments with transgenic crops aimed to enhance plant resistance to insect pests. While gene transfer was becoming routine in the late 1980s, the identification of useful genes for transferring to crops was advancing at a slower rate. Several genes coding for different types of insect toxins had been identified, however, which were used to develop insect-resistant transgenic crops. These included genes from the bacterium *Bacillus thuringiensis*, and genes from plants in the family Leguminosae that expressed insect toxins.

Bacillus thuringiensis toxin

Bacillus thuringiensis (*B.t.*) is a soil bacterium that accumulates high levels of insecticidal proteins during sporulation, when bacterial cells transform themselves into spores. Bacterial spores are formed in order to survive adverse environmental conditions. They can remain dormant for considerable periods of time in the soil, before the bacteria's life-cycle is resumed. The toxin proteins can amount to almost 20 per cent of the weight of the bacterial spore. When insect larvae ingest the bacterial spores, the spores dissolve in the highly alkaline gut to release the toxins. These toxins bind to the membrane of the gut walls, paralysing the gut and preventing nutrient uptake. The insects stop feeding and die. *B.t.* toxins are highly specific to particular groups of insects, kill only larval stages, and are not toxic to other organisms. They have been used as commercial insecticides since 1958, with spray formulations generated by fermenting the spores. They are biodegradable and safe for humans and non-target organisms, and are therefore a preferred option for use in environmentally sensitive applications. In the final chapter of her classic book *Silent Spring*, first published in 1962, Rachel Carson saw *B.t.* sprays, along with biological control, as the way forward from persistent and environmentally damaging insecticides such as DDT. Production has increased considerably since the 1960s. *B.t.* accounts for most of the biopesticide market, and sales are forecast to be worth around US$300 million by the end of the twentieth century.[1] The use of *B.t.* has, however, been constrained by high production costs and poor persistence in the field, due to the instability of the crystal proteins.

The first DNA sequence of a gene coding for *B.t.* toxin was obtained in 1985. A great diversity of *B.t.* strains from which to select toxin genes occur in nature. The company Mycogen, for instance, has collected several thousand strains for screening, from around fifty countries.[1] A large number of *B.t.* toxin genes have now been cloned and sequenced. They can be grouped into four major classes: class *cryI* genes are the most thoroughly studied and are highly specific to species of moths and butterflies (lepidoptera); *cryII* genes have a complex activity spectrum, against various lepidoptera, flies (diptera) and beetles (coleoptera); *cryIII* genes are active against coleoptera; and *cryIV* genes are active against diptera.[2] *B.t.* toxins are the products of single genes. In addition, their safety, efficacy and the relative simplicity of their structure and genetics made them ideal for early research on transgenic crop development.[3]

Genes expressing *B.t.* toxins were first engineered into a crop plant, tobacco (*Nicotiana tabacum*), using *Agrobacterium*, by the Belgian-based company Plant Genetic Systems.[4] Tobacco, a relative of potato (*Solanum tuberosum*) and tomato (*Lycopersicon esculentum*), is a standard experimental plant for this type of research. Leaves of transgenic tobacco were highly toxic to tobacco hornworm (*Manduca sexta*), an important economic pest of tobacco. The progeny, grown from seeds of these plants, were also resistant. However, caterpillars of other economically important moths, for example species of *Heliothis* and *Spodoptera*, proved to be less sensitive to the *B.t.* toxin and would therefore require higher levels of expression of transformed genes in transgenic plants.[4] Monsanto reported similar results later in the summer of 1987.[5] Soon after this, the Agracetus Company reported success with transgenic tomato plants, again obtained using an *Agrobacterium*-mediated transfer of a *B.t.* toxin gene, which conferred resistance to lepidopteran larvae on plants and the progeny of those plants.[6] In the USA, the first approval for a field test of plants (tobacco) containing a *B.t.* gene was given as early as 1986.[7] The first commercial crops that were genetically engineered to produce an insecticide, containing a gene expressing *B.t.* toxin, were given final approval by the US Environmental Protection Agency (EPA) in 1995. These included a genetically modified Russet Burbank, the most popular US potato variety, which was produced by Monsanto in collaboration with the University of Wisconsin. These NewLeaf™ potatoes were resistant to the Colorado potato beetle (*Leptinotarsa decemlineata*).

A frequent problem has been obtaining high levels of *B.t.* toxin gene expression in transgenic plants. In one pilot study, a toxin produced by an introduced gene was inactivated by the ultraviolet component of sunlight.[8] However, this and other problems are gradually being overcome. All the major agrochemical and biotechnology companies are developing transgenic crop plants incorporating *B.t.* toxin genes.[1]

Ciba-Geigy's *B.t.* maize

Maize (*Zea mays*) has been modified for resistance to the European corn borer (*Ostrinia nubilalis*) and other insects by a number of research groups. For example, Monsanto's *B.t.* toxin gene YieldGard™ is incorporated into Golden Harvest Seeds' hybrid maize seed.[9] The European corn borer can infest about 24 million hectares in the USA and can cause yield losses of up to 20 per cent of the total crop. Large quantities of insecticides, amounting to US$20–30 million, are used against this pest each year, but it is difficult to control because the insect spends much of its life-cycle inside the plant.

A transgenic maize developed by Ciba-Geigy, now part of the Swiss-based multinational Novartis,[10] incorporated a gene expressing *B.t.* toxin against the European corn borer. It also incorporated a gene expressing an enzyme that confers resistance to the herbicide Basta™, whose active ingredient is glufosinate ammonium (see Chapter 4).[11] As with many transgenic plants, it also incorporated a selectable marker gene conferring antibiotic-resistance, in this case against ampicillin, and promoter genes, to control foreign gene expression in the transgenic organism. Ciba-Geigy/Novartis' Maximizer™ maize was the first commercial transgenic crop to incorporate both insect and herbicide resistance characteristics.

The *B.t.* toxin gene was integrated into the maize in a process involving several stages and two marker genes. The *B.t.* gene was first linked with a marker gene and integrated into a plasmid of a bacterial host. The antibiotic marker gene, originating from the bacterium *Salmonella parathypi* and called the *bla* gene, causes the production of an enzyme (TEM1-lactamase) that inactivates ampicillin. The marker gene is under the control of a bacterial promoter gene and is present, but not expressed, in maize. An antibiotic treatment selects for bacterial cells containing the ampicillin resistance gene. These cells also contain the *B.t.* toxin gene, which is closely linked to the antibiotic resistance gene due to its proximity in the vector. These selected bacteria were allowed to reproduce, and then their plasmids were isolated.[11]

The gene conferring resistance to the herbicide glufosinate ammonium, called the *bar* gene, was initially isolated from a plasmid of the bacterium *Streptomyces hygroscopius*. This gene, in the presence of promoter genes, expresses an enzyme (phosphinothricin acetyl transferase) that confers tolerance to glufosinate ammonium, by overproducing the enzyme that the herbicide inhibits to achieve its toxic effect. The gene was integrated into a bacterial plasmid and the bacteria cloned. The plasmids were then isolated.[11]

The two types of plasmid, containing the *B.t.* toxin and the herbicide resistance genes, were then fired simultaneously at maize plant cells using a particle gun. Maize cells were grown in nutrient medium and then sprayed with the herbicide Basta™ (glufosinate ammonium). The surviving and reproducing cells were those that had incorporated the herbicide resistance gene. Some of these cells also carried the *B.t.* toxin. The plants grown from transformed cells were then crossed with other varieties using traditional plant-breeding techniques to produce commercial hybrid seed.[11]

Ciba Seeds, who marketed the hybrid transgenic seed in the mid-1990s, claimed that the Basta™ resistance gene was being used only as a development tool.[11] The use of Basta™ was not allowed on maize at the time Ciba's maize reached the market. Applications have since been made by Hoechst/AgrEvo, however, for the use of Basta™ on maize. The success of these applications would prove mutually beneficial to both companies, with increased sales of herbicide and the selling of premium-priced herbicide-resistant transgenic seed. Maize is known to be sensitive to glufosinate ammonium and other herbicides in the phosphinothricin group, which is a major constraint on their use in this crop. The development of maize varieties resistant to glufosinate ammonium is, therefore, likely to be greeted enthusiastically by growers with weed problems in maize, particularly as they also contain a gene effective against the European corn borer.

In 1996, Ciba-Geigy reported that its transgenic maize seed was sold out within a couple of days. In 1996, it was planted on 18,000 hectares in the USA, amounting to 0.61 per cent of the year's total crop.[11] In 1997, it accounted for around 8 per cent of the total maize crop. This figure is likely to increase further.

Protease inhibitors and lectins

Particular *Bacillus thuringiensis* toxins are specific only against certain groups of insect pests. This is advantageous in many respects, but it means that a particular *B.t.* toxin gene will have limited crop protection applications. A number of more general insect-resistance mechanisms exist, however, that confer tolerance to a wide range of insect pest species. Protease inhibitors, for example, are widely distributed within the plant kingdom, particularly in seeds and storage organs, where they usually account for between 1 and 10 per cent of protein content. These molecules form complexes with particular animal or microbial digestive enzymes, preventing them from breaking down proteins. They therefore play an important defensive role against predators.[12] Protease inhibitor genes are often expressed in response to insect attack or other damage, causing inhibitors to accumulate in the foliage after wounding. Chemicals called protease inhibitor inducing factors (PIIFs) are released when plant tissue is damaged and these induce the formation of protease inhibitors, a response that can occur within ten seconds and last for several hours.[13] The protease inhibitors are transported around the plant in the vascular system. Some reports suggest that wound-induced defences can even be provoked in plants near to those that are attacked.

Genes expressing protease inhibitors were first identified from potato by Clarence Ryan's group at Washington State University, USA.[13] A number of distinct families of protease inhibitors have been identified in plant tissue. Research has concentrated on one of these families, the trypsin inhibitors from cowpea (*Vigna unguiculata*), which are known to contribute to resistance against the cowpea seed beetle (*Callosobruchus maculatus*) and other insect pests in the field.[14] Cowpea trypsin inhibitors (CpTIs) act on a catalytic site of a protease enzyme called trypsin, preventing insects from digesting proteins in their diet. Purified trypsin inhibitors, presented in artificial diets, have shown anti-metabolic effects against a wide range of pest insects. A gene coding for a CpTI was subsequently identified and transferred to leaf disks of tobacco using *Agrobacterium tumefaciens*.[15] The expression of the CpTI gene resulted in enhanced resistance to the tobacco budworm (*Heliothis virescens*). This work was done by the Agricultural Genetics Company, based in Cambridge, England, which holds the proprietary rights to CpTI.

High levels of protease inhibitors need to be expressed in leaves to ensure that they disrupt insect digestion. Transgenic plants containing protease inhibitor genes are more likely to slow down the development of a pest population than to eliminate insects, although mortality will be increased to a certain extent. Therefore, selection pressure for insects to develop resistance to protease inhibitors will be less than for *B.t.* and other acute toxins. However, this resistance mechanism might make them harder to sell to growers, compared to methods giving more tangible evidence of pest mortality.

Lectins are a group of plant-derived proteins that cause cells to group together (agglutinate). The lectins found in the seeds and tissues of legumes are toxic to insects that are not adapted to living on plants in the family Leguminosae. The pea's lectin gene[16] has been transferred to potatoes to confer resistance to Colorado potato beetle and other pests. A field of Desiree potatoes containing a lectin gene was grown in 1990, from engineered tubers, by Nickerson International in Norfolk. This was the first major planting of a genetically modified crop in the UK.[17] Lectin genes have also been identified from other plant families. A gene from snowdrops (*Galanthus nivalis*), a plant in the family Liliacae, has been transferred to potatoes and tested for insecticidal activity by Axis Genetics, based in Cambridge, England. The snowdrop lectin was found to have a repellent and antifeedant effect against aphids. Field trials with transgenic potatoes containing the lectin gene were carried out at Rothamsted Experimental Station in 1995 and 1996.[18] Genes expressing protease inhibitors and lectins have also been used in combination with other genes, for example *B.t.* toxin genes, in transgenic plants.

Pyramiding genes

The packaging of several different genes, expressing proteins with different functions, into a vector to produce a multi-gene 'cassette' has opened up many interesting possibilities for transgenic plant development. Combining different characters, a process sometimes called 'pyramiding', into the same plant is a strategy based on what plants do naturally to protect themselves. This approach was first achieved in practice using genes for different mechanisms of insect resistance.[19] Genes coding for a protease inhibitor and a pea lectin were integrated into tobacco. The transgenic plant showed additive effects of the two gene products against tobacco budworm (*Heliothis virescens*). Therefore, this approach may provide plants with better protection against pest insects. Pyramiding genes should also reduce the ability of insects to develop resistance to insecticidal toxins expressed in transgenic plants.

A variety of transgenic cotton, incorporating a gene expressing a *B.t.* toxin and a gene expressing high levels of terpenes (chemicals found in the essential oils of plants) was developed by Monsanto with the intention of slowing the development of insect resistance to *B.t.* toxins. The cotton had an increased level of resistance to the tobacco budworm, compared to plants expressing these two insect resistance characters separately.[20] Genes expressing resistance to any new character are likely to be initially rare and to arise independently of any other resistance mechanisms. Therefore, individual insects with genes responsible for resistance to two or more pyramided toxins in a transgenic plant will be exceedingly rare. To be most effective, however, the pyramiding approach needs to be deployed early, before resistance starts to develop to any one character. Protease inhibitors, with their general anti-herbivore effect of inhibiting protein digestion, have been proposed as ideal candidates for multi-gene packages, alongside genes expressing pest-specific toxins.[14] Pyramiding of genes could be extended to any food crop, where two or more characters confer a particular resistance advantage.

The vector constructs used to produce transgenic plants are likely to become more complex in the future. As well as carrying pyramiding genes that express particular proteins, these constructs will contain an increasing number of regulatory genes, including transcription promoters and enhancer and silencer sequences. Different promoters could be used for crops growing under different environmental conditions, to modify the expression of a gene. Genes could also be regulated so that they are expressed only under certain conditions. Genes coding for insect toxins, for instance, could be regulated by genes that ensure the toxin is expressed only at certain times or in certain situations – for instance, in green tissue only.

The production of transgenic plants with multiple genes is likely to become an increasingly common method of engineering crops. Libraries of genes for particular characters, and their associated regulatory genes, are being constructed by the major multinationals. As more genes are patented, cross-licensing agreements between companies will enable diverse characters to be incorporated into the same transgenic varieties. For example, an agreement between Monsanto and Calgene led to the integration of genes expressing speciality oils in rape crops to be accompanied by genes expressing herbicide resistance.[21] A number of patented *B.t.* toxin genes, active against different insect groups, are now available for 'plugging into' crops, with royalties payable to the licence owners. A range of 'designer plants', containing proprietary genes for a number of different characters, may soon be on the market.

A practical upper limit to the number of genes incorporated into transgenic plants may result from the potential difficulties encountered by manufacturers of food products in getting approval for marketing. A greater number of gene modifications amplifies the number of factors that have to be analysed during risk assessment. Critics of Ciba-Geigy/Novartis' maize were able to raise concerns about insect toxins, increased herbicide use during crop production and the possibility of antibiotic resistance transferring to micro-organisms living in the human gut. The increasing number of variants of genetically modified crops, with different gene combinations, may mean that different variants have different markers, whose identity is preserved from the farm through processing into food items. The particular combination of characters may effect a crop's suitability for certain food-related uses.

One of the major constraints on the genetic engineering of plants is that many important characters are controlled by multiple genes – that is, many genes are needed to give one desirable character. Current techniques are most effective when dealing with characters controlled by a single gene, such as *B.t.* toxins. Complicated proteins could, however, be produced in stages. The immunoglobin antibody, consisting of two chains, has been produced in tobacco by crossing two transgenic tobacco varieties, each carrying a different chain of the protein. A functional two-chain antibody was present in the offspring.[22] This technique might also expand the possibilities for transgenic crops grown for food production.

Benefits for insect control

The genetic modification of crops offers many potential advantages over existing insect pest control methods.[3] Growers are no longer dependent on the weather, as the crop will be protected even when the weather is too severe, or fields too muddy, for conventional spraying. Also, plant parts that are difficult to reach by spraying, such as lower leaves and roots, and new growth emerging between spray applications, would be protected, and as the control agent is continuously present in the field, there is no need for scouting to determine spray timing. In crops with heavy insecticide applications, insect-resistant transgenic crops could bring large economic benefits to growers, with savings made on insecticide expenditures, labour and equipment.

The cost of developing a genetically engineered seed variety represents a much smaller investment than producing a new chemical insecticide. The cost of inventing, developing, registering and producing a new chemical insecticide is well over US$25 million, while the cost of developing a new crop variety may be under US$1 million.[3]

The use of insect-resistant crops also promises improvements for the environment compared to conventional insecticide spraying. Less insecticide will be needed on insect-resistant transgenic crops, as the insecticidal material is contained in the plant tissue. Therefore, spray drift and its associated problems will be less of a concern. Contamination of groundwater around farmland should also be significantly reduced. With no insecticide spraying, the deleterious effects on beneficial insects will be alleviated, leading to potentially more effective biological control of pest insects, acting in conjunction with the effects of the toxin. Other non-target organisms, from bees and earthworms to birds and mammals, will not be exposed to insecticide through spraying. The *B.t.* toxin and trypsin inhibitor compounds are natural proteins, which are readily biodegradable into non-toxic substances.

Beneficial effects for human health have also been proposed for engineered crops compared to conventionally sprayed crops. The number of spray operator poisoning incidents will be reduced, an effect that may be most beneficial in developing countries where adequate training and protective equipment is often lacking. Monitoring for safety for human consumption should be easier with transgenic crops compared to those where chemicals are applied. The nature of the added materials is known in advance for transgenic crops, with the foreign genes being fully characterized.[3] Assessing risks from conventional spray residues, however, involves the use of expensive analysis equipment. The toxicology of many conventional spray residues is complex, because the cocktail of chemicals may consist, for example, of more than one insecticide, a herbicide and a fungicide. These mixtures often involve substances potentially injurious to human health, and unforeseen interactions between chemicals sometimes occur. Overall, pesticide residues, for example in vegetables, should decline with reduced insecticide use. In traditional plant breeding, the effects of insecticidal products, in varieties bred for increased insect resistance, are often unknown. Traditional plant breeding is largely unregulated, while the effects of engineered crops are closely monitored and regulated.

Resistance management

However, a major concern in the deployment of crops engineered with insect toxins is the potential development of resistance by insects to the toxins. This situation mimics the natural evolutionary 'arms race' between insects and plants: plants evolving new defences escape insect attack, until the insects evolve counter-adaptations. As with the development of resistance to conventional insecticides, an individual insect with a genetic predisposition or a mutation that enables it to survive on a toxin-containing plant will be selected for in preference to susceptible insects, causing the resistance gene to spread in the population. The partial failure of transgenic crops to control insects may speed up the evolution of insect resistance.

Resistance to *B.t.* sprays, which are increasingly used by organic farmers because of their high specificity to pest insects, has developed in the field. This occurred about a decade after their use started.[23] The situation could worsen if transgenic crops containing genes coding for *B.t.* toxins become widespread. The cumulative impact of an array of different crops containing *B.t.* genes could raise pest resistance to a point where *B.t.* spraying soon becomes ineffective. This could have severe consequences for biological control programmes that incorporate *B.t.* spray regimes. Resistance to *B.t.* will develop much faster using transgenic plants than using sprays, because plants produce the toxin continuously. An additional concern, reported in 1997, was that a single gene in the diamondback moth (*Plutella xylostella*), a major pest of brassica crops, conferred resistance to four different *B.t.* toxins. It had previously been thought that changing the type of *B.t.* toxin would slow the spread of resistance, but the development of cross-resistance to *B.t.* toxins is more common than previously thought.[24]

A key part of resistance management is the establishment of 'refugia', areas in a field or on a farm where plants are grown that do not contain, for example, *B.t.* toxins. Pest populations that are not resistant to toxins are maintained on these refugia plants. If whole areas were to be grown with transgenic *B.t.* crops, few susceptible insects would survive, and insects resistant to *B.t.* would quickly dominate the population. By mating with resistant insects, refugia insects dilute the effects of insect genes that confer *B.t.* resistance. Monsanto, for example, recommends the use of refugia to growers of its Bollgard™ *B.t.* cotton.[25] In 1996, the majority of growers used refugia. The recommended resistance management strategy for Monsanto's NewLeaf™ *B.t.* potato, which is marketed by its subsidiary NatureMark, is to avoid planting the *B.t.* potato on all the fields within a farm in any one year, or on any one field two years in succession. A knowledge of regional variation in pest insects and their wild and cultivated host plants will be necessary for adapting resistance management strategies to particular localities. A close monitoring for the presence of resistant pest insects is also necessary. Resistance management is therefore a vital, and complex, component in the long-term use of transgenic crops modified with genes expressing insect toxins. The size of the refugia required for resistance management is, however, a contentious issue. It is also uncertain whether the toxin is being delivered in doses that are high enough for the refugia strategy to work in current *B.t.* crops. Expression of *B.t.* often tapers off during the growth season and can sometimes go below a lethal dose, allowing increased insect survival and speeding up the evolution of resistance. Larger refugia and higher expression of toxin, for example, in Bollgard™ cotton may be needed for refugia to be most effective.[26]

The ability of insects to develop resistance to the other toxins present in transgenic crops may be of less immediate concern. The metabolic target of trypsin inhibitors is the catalytic site of an enzyme, and therefore the ability of insects to evolve a resistance mechanism to these toxins should be less than for *B.t.* toxins.[15] Lectins also have a mode of action that will be more difficult to evolve resistance against. However, in the long term, resistance management strategies will need to be operating to safeguard the efficient use of all transgenic varieties that exert selection pressure on pest insects.

An argument in favour of moving so quickly to the commercial cultivation of transgenic crops was that it was the only way to understand fully how best to manage resistance.[27] Large-scale experiments were done, therefore, to monitor commercially grown crops for resistance development. This was a controversial decision and critics saw these commercial plantings as premature. They argued that approving the release of pesticidal plants, with a fairly clear knowledge that they will hasten resistance development, was not a responsible way to pursue a long-term pest control policy.[27] Commercial pressures for moving quickly to large-scale plantings of transgenic crops can mean that basic insect–plant interaction studies are often not adequately completed. Further information, for example correlating the mortality levels of pests to toxin levels present in leaves, would slow down the deployment of insect-resistant crops, not something that companies wanting a return on their investment want. However, this lack of information may ultimately result in crop failures, which may be detrimental to the companies involved.

Problems with commercially grown transgenic crops were reported in both 1996 and 1997, particularly with transgenic cotton.[25] In 1996, Monsanto's *B.t.* Bollgard™ cotton failed to protect against cotton bollworm (*Pectinophora gossypiella*) and other insects it was designed to kill. The transgenic cotton was planted by 5,700 growers on around 0.8 million hectares, accounting for about 13 per cent of the total US cotton crop for that year. Eight thousand hectares of this cotton were devastated by insect attack, with losses estimated at over US$1 billion. The leaves of the cotton were subsequently shown to be producing too little toxin to be detrimental to the caterpillars. This may be because toxins were being expressed at too low a level, or because toxins were not uniformly distributed in the leaves, so that insects were avoiding areas of high toxin in a 'toxic mosaic' and feeding on leaf areas with low toxicity.[26] Monsanto advised farmers to spray conventional insecticides to try and save as much of the cotton crop as possible. The hype behind Bollgard™ *B.t.* cotton may have given farmers unrealistically high expectations of how it would cope with high population levels of cotton bollworm. Monsanto maintained that the problem had not damaged farmer confidence, however, as a survey of these cotton growers after the 1996 harvest found the majority to be satisfied with the product's performance. Monsanto reported that growers gained an economic advantage of about US$33 per hectare from using Bollgard™ cotton, while the use of insecticide sprays was significantly down. However, critics argued that reductions in insecticides are going to occur only in the short term, as pests will soon become resistant to *B.t.* toxins in transgenic plants, resulting in the spraying of more expensive and hazardous insecticides to combat pest problems. In 1997, about one quarter of the total US cotton crop was grown using transgenic varieties and soon a majority of the crop may be transgenic, making sound resistance management of vital importance if the benefits of *B.t* cotton are going to be of any lasting value.

There were further problems with Monsanto's transgenic cotton, however, in 1997. Crop failures occurred in Mississippi, while nearby non-transgenic varieties were unaffected. It is suspected that cold weather affected the transgenic varieties, leading to boll damage. More than forty farmers have filed complaints against Monsanto and are seeking millions of dollars in damages.

Baculovirus: engineering a quicker kill

Another approach to combating insect pests is to insert genes into organisms that are naturally insecticidal, to make them more effective as insect pathogens. The baculoviruses, or nuclear polyhedrosis viruses, cause disease in the larval stages of a few species of insects.[28] The infected insects are referred to as the 'permissive hosts'. One of these baculoviruses is the *Autographa californica* nuclear polyhedrosis virus (AcNPV), which naturally infects the alfalfa looper moth (*Autographa californica*) and a range of related moth species. AcNPV is a registered insecticide in the USA, where it has been tested for safety under the protocols of the Environmental Protection Agency (EPA). Like other baculoviruses, AcNPV has many desirable features as an insecticide: it infects only arthropods, not vertebrates or plants, while its host range is limited to a small number of moth families. Baculoviruses also have good storage properties, are safe to handle and are relatively easy to produce. However, a major constraint on their use is the slow action of these agents, which take at least three to five days to kill their permissive hosts. Insect crop pests can continue to consume plants while the baculovirus is taking its effect. Genetic engineers have aimed to incorporate a quicker-acting toxin within the baculovirus to reduce the time needed to achieve mortality of insects, and therefore reduce the damage to crops. A range of genes coding for insect neurohormones or insect-specific toxins have been integrated into baculovirus.[29] In 1991, toxin genes from the mite *Pyemotes tritici* and the North African Scorpion *Androctonus australis* were successfully transferred to baculovirus.[30] The latter was to prove the most promising for insect control purposes.

Research on genetically modified baculovirus started at the National Environmental Research Council's (NERC) Institute of Virology and Environmental Microbiology in Oxford, England, in 1986. In contrast to the secrecy surrounding much commercially sponsored research, the details of this research were readily made available to the public. In a letter to the journal *Nature*, the institute's director, David Bishop, informed the scientific community of the first experimental release of modified AcNPV at Oxford's Wytham Field Station, prior to publication of a full analysis of the data.[31] The AcNPV was modified to include a genetic marker, which did not affect the viral gene product, but allowed the micro-organisms to be tracked in the environment. The modified baculovirus was released not by spraying, but within infected caterpillars of the small mottled willow moth (*Spodoptera exigua*). The purpose of this release was to study the ability of the marked baculovirus to exist in the ecosystem, prior to releases of baculovirus modified with insect toxin genes.

In Oxford, work proceeded with AcNPV modified to express an insect-selective toxin gene derived from scorpion venom. In laboratory assays with this modified AcNPV, a 25 per cent reduction in the time to death of the cabbage looper (*Trichoplusia ni*) was observed, compared to wild type baculovirus. The first field trials showed that the modified baculovirus killed caterpillars faster, but to a less dramatic extent than in the laboratory, with reduction in time to kill from 7.3 to 6.2 days, for the highest dose rate, compared to wild type virus.[32] The reduced time to kill, however, resulted in significantly reduced crop damage. Modified virus also resulted in a reduced secondary cycle of infection, because fewer caterpillars lived long enough to release further baculovirus to the environment, with the result that modified virus-infected hosts released only 10 per cent as many virus particles as wild-type hosts. In addition, caterpillars infected with altered baculovirus are immediately paralysed and fall to the ground, taking them out of the reach of other feeding caterpillars, and the caterpillar bodies stay intact, as too few viruses have multiplied within them to liquefy their bodies as happens with natural infection, so viruses do not escape.[33] Baculoviruses are inert outside host cells, and eventually degrade in the soil. All these reasons, along with the narrow host range of baculovirus, led the NERC researchers to conclude that modified baculovirus is unlikely to spread or persist in the environment.

Meanwhile, this work became the focus of debate on the potential of modified viruses to affect non-target species. Critics argued that the altered virus might escape to infect other species of moth and that the scorpion gene might cross into other types of virus, making them more dangerous. Some of the strongest critics, including those within the Zoology Department of Oxford University, were alarmed that field trials were being done close to Wytham Wood, an important ecological study area and nature reserve, where a large number of native lepidoptera could potentially become exposed to modified baculovirus. A host range study had focused on the British lepidoptera, using 58 moth and 17 butterfly species, showing no difference in host range between modified and wild-type virus.[34] Although some species, such as the lime hawkmoth (*Mimas tiliae*) and the privet hawkmoth (*Sphynx ligustri*) were found to be permissive, high doses were required for them to become infected. However, as Mark Williamson pointed out in a letter to *Nature*,[35] 5 to 10 per cent of British lepidoptera are potentially permissive for AcNPV, which is a non-native baculovirus, putting at potential risk 125 to 250 species, including some of great conservation value. Bruce Hammock,[36] in response, pointed out that the baculovirus was designed as an insecticide that disappears rapidly from the environment, not as a biological control agent that becomes permanently established, and that it was intrinsically more species-specific than either synthetic chemical insecticides or *B.t.* sprays.

The view of genetically modified baculovirus as being analogous to classical insecticides is, however, probably not shared by the majority of the population. Public unease about the virus experiments during the summer of 1994, when modified baculovirus was sprayed on cabbage plots under field trial conditions, prompted David Bishop to call a public meeting in Oxford in November of that year to explain his group's results. The public were reassured that the chance of baculovirus escaping and establishing in the wider environment was extremely small. All the field trials at Oxford have been carried out under strict release conditions, involving containment in field cages, and in consultation with a range of organizations and government departments, including the Nature Conservancy and the Department of the Environment.[37] In fact, one of the major contributions of the baculovirus project has been its input into the development of protocols now used for the release of all genetically modified organisms into the environment.

In January 1995, David Bishop told a meeting of the Royal Entomological Society, in London, that genetically engineered baculovirus would need to be better than the scorpion gene AcPNV to be of real benefit to farmers.[38] The modified baculovirus was still taking the best part of a week to kill caterpillars, compared to hours for conventional insecticides. Various approaches could make this model system more effective, including transferring more copies of the scorpion toxin gene and making the toxin more stable within the host insect. The Oxford entomologist George McGovern expressed reservations about making the baculovirus a more efficient killer, because it could potentially infect a wider range of species. Terry Tooby, the deputy director of the Ministry of Agriculture's Pesticides Safety Directorate, also showed concern about a toxin being stabilized, saying: 'Just because it is possible to carry out some of these ideas, it does not mean it is advisable.'[38]

In March 1995, David Bishop was removed from his job as director of the Institute of Virology and Environmental Microbiology.[39] The management at NERC decided that the institute needed a new mission and new leadership. Another public meeting, in Oxford in November 1995, saw the remaining members of the team explaining their latest field work results. The emphasis was on the assessment of risks and safety. A series of field trials is being completed to assess the host range of baculovirus and the survival of caterpillars under different environmental conditions.[40] Local protests against the work have continued.[41] The initial funding ends in 1998 and the future of the project is uncertain.

Experimental releases of modified baculovirus have been relatively few elsewhere in Europe.[42] In the USA, a higher number of baculovirus studies have been conducted. In January 1995, however, the EPA withheld approval for an experimental release of a baculovirus containing a scorpion toxin gene in the USA, which had been proposed by American Cyanamid. This organism was very similar to the one used in the UK field trials. The main concern of the EPA was the potential effect it could have on non-target organisms. However, in September 1996 American Cyanamid were given approval by the EPA for field releases in twelve states to test the efficacy of a baculovirus incorporating a scorpion toxin gene (AaIT strain) against tobacco budworm and cabbage looper on cotton, tobacco, cabbage, broccoli and lettuce. On this occasion the EPA concluded that the baculovirus posed no significant risk to human health or non-target organisms. These releases are, of course, not without their critics. Concern was expressed about this modified AaIT baculovirus strain having a rate of infection greater than the unmodified baculovirus. The experimental procedures were also questioned, particularly the widespread use of lime as a method of inactivating baculovirus after the field trials.[43] In general, regulations covering engineered biopesticides are less strict in the USA, compared to the UK, and the first modified baculovirus could be marketed to American maize and cotton growers by 1999.

Notes

1. Feitelson, Payne and Kim, 1992.
2. Barton and Miller, 1993.
3. Meeusen and Warren, 1989.
4. Vaeck et al., 1987.
5. Barton et al., 1987.
6. Fischhoff et al., 1987.
7. *Chemical Week*, 139: 34, 1986.
8. Lal and Lal, 1990.
9. *AgBiotech News and Information* 8 (12): 203N, 1996. Monsanto subsequently bought the Golden Harvest seed group in 1997.
10. Ciba-Geigy merged with Sandoz in 1996 to form a giant new Swiss company called Novartis. Ciba Seeds was compartmentalized within the parent company Ciba-Geigy when *B.t.* maize was first marketed. The herbicide Basta™ was originally developed by Hoescht/AgrEvo. Ciba Seeds claimed that it 'neither produce[s] or sell[s] Basta herbicide'. However, its new parent company Novartis does manufacture and sell Basta™.
11. Documentation on *Bt*-maize from Ciba Seeds. 1996. Ciba Seeds, Basel, Switzerland.
12. Gatehouse and Boulter, 1983.
13. Ryan, 1990.
14. Hilder et al., 1993.
15. Hilder et al., 1987.
16. Gatehouse et al., 1987.
17. *New Scientist*, 8 September 1990, p. 43.
18. Department of the Environment (UK), Public Register of Genetically Modified Organism Releases: 95/R17/2. A harmful effect of lectins on natural predators of aphids was reported in late 1997. Ladybirds fed on aphids that had fed on potatoes engineered with a lectin gene from snowdrops lived only half as long as ladybirds fed on aphids from normal plants. In addition, fewer viable eggs were laid by ladybirds fed on aphids from lectin-engineered potatoes. *New Scientist*, 1 November 1997, p. 5.
19. Boulter et al., 1990.
20. Sachs et al., 1997.
21. *AgBiotech: News and Information* 8 (9): 155–6, September 1996.
22. Hiatt, Cafferkey and Bowdish, 1989.
23. Tabashnik, 1994.
24. Tabashnik et al., 1997.
25. *Science* 273: 1641, 20 September 1996; *Nature Biotechnology* 15: 1233, November 1997.
26. *Science* 273: 423, 26 July 1996.
27. *New Scientist*, 6 May 1995, p. 9.

28. The baculoviruses or nuclear polyhedrosis viruses are members of the Eubaculovirinae, a sub-family of the family Baculoviridae.
29. Wood, 1995.
30. Tomalski and Miller, 1991; Stewart et al., 1991.
31. *Nature* 323: 496, 9 October 1986.
32. Cory et al., 1994.
33. *New Scientist*, 3 December 1994, p. 11.
34. Bishop et al., 1988.
35. *Nature* 353: 394, 3 October 1991.
36. *Nature* 355: 119, 9 January 1992.
37. Ager, 1988.
38. *New Scientist*, 21 January 1995, p. 6.
39. *New Scientist*, 25 March 1995, p. 6.
40. Department of the Environment (UK), Public Registry of Genetically Modified Organism Releases: 93/R3/4.
41. *New Scientist*, 25 November 1995, p. 8.
42. Landsmann and Shah, 1995.
43. *AgBiotech: News and Information* 8 (9): 158N, 1996.

6. Designer food and engineered plants

The potential of genetic engineering to modify crops is enormous. In this chapter, a wide range of crop improvements that have been developed, or are in the process of being developed, are described.

Modifications for food processing and taste

The first genetically engineered vegetables to reach the market were tomatoes. Tomatoes have a relatively small genome, are easy to work with and were used in some of the early experimental work on genetic manipulation techniques. Two research groups pioneered the genetic modification of tomatoes for delayed ripening: Calgene in the USA, and the University of Nottingham, in collaboration with Zeneca (then a division in ICI), in the UK. Both groups used gene-silencing technology.

Calgene, based in Davis, California,[1] were commissioned by Campbell's Soups, who process hundreds of thousands of tons of tomatoes every year, to develop a tomato that stays firm for longer. A South American mutant tomato that never ripened provided the clues for engineering tomatoes that resist the decaying process. The gene that makes tomatoes go soft during the ripening process was first identified. This gene expressed an enzyme called polygalacturonase (PG), one of a group of enzymes called pectinases. These enzymes break down pectin, a major component of cell walls, resulting in the conversion of solid plant tissue into softer tissue during the ripening process. The Calgene group then produced an antisense gene, an artificially manufactured DNA sequence containing bases complementary to the target PG gene's DNA sequence. This antisense gene was transferred, along with a promoter gene and an antibiotic marker gene, into a tomato genome. The antisense gene's mRNA binds to, or hybridizes with, the mRNA from the endogenous PG gene, before it can be expressed, and inactivates it.[2] By preventing the expression of the PG gene, no pectinase enzyme is produced and the softening process does not occur, although other ripening processes such as flavour and colour changes are unaffected. Calgene's tomatoes produced only 1 per cent of the normal amount of the enzyme PG. Normally, as a fruit softens, more bacteria and fungi are attracted, which hastens the rotting process. Firmer tomatoes therefore have a higher solid to water ratio and are very slow to rot.

In commercial terms, tomatoes thus modified could be vine ripened and still have a substantial shelf-life. Vine-ripened tomatoes generally have a better taste than tomatoes picked and ripened in transit, and retain their flavour and texture longer. Because of their flavour properties these modified tomatoes were called Flavr Savr™, the name Calgene gave to the patented antisense gene used in their production. However, the most important factor for the food processors was the reduced water content. The higher solid to water ratio of the engineered tomatoes at the time of picking meant that big savings could be made in harvesting, trucking and processing costs, as tomatoes are more concentrated when processed to paste. Other potential benefits included reductions in damage to produce in transit and fewer fungicide applications. These tomatoes were initially used in Campbell's Soups' soups, paste and ketchup.[3]

In the USA, Flavr Savr™ tomatoes were cleared for sale by the Food and Drug Administration (FDA) in May 1994. They became the first fresh genetically modified fruit or vegetable to reach the market, when they went on sale in late 1994 in 730 stores across the USA. A detailed report on Flavr Savr™ tomatoes by the FDA showed no nutritional differences between them and conventional tomatoes. However, the report noted that the tomatoes contained a gene that expressed a protein conferring resistance to the antibiotics kanamycin and neomycin. This marker gene was used to select transformed plant material during the modification process. Consumers would have been unaware of this, or any potential risks of antibiotic genes, at this time.

The initial sales of the Flavr Savr™ were promising and now Calgene's tomatoes are marketed under the McGregor brand name and sold in over three thousand stores throughout the western states of the USA. Permission was granted in 1995 to sell Calgene's genetically modified tomatoes in Mexico and Canada. However, during the scaling-up operation for the fresh produce market, low yields were obtained due to production problems with the original Flavr Savr™ tomatoes. The growing operations were curtailed while new varieties were developed with better growth characteristics, by using traditional plant-breeding methods to cross plants containing the Flavr Savr™ gene with other varieties. By late 1996, Monsanto owned a majority share in Calgene.[4]

Calgene was given safety clearance by the UK government in February 1996 to market ten lines of genetically modified tomatoes containing the Flavr Savr™ gene.[5] This was the first clearance of an unprocessed genetically modified food anywhere in Europe. The government's Advisory Committee on Novel Foods and Processes (ACNFP) decided that the presence of the antibiotic marker genes would not compromise the clinical or veterinary uses of antibiotics. However, the tomatoes are unlikely to be on sale in Britain until at least 1998, as Calgene has to wait for EU approval on a submitted application before they can market the tomatoes in Europe.

The research collaboration between Don Grierson's team at the University of Nottingham and Zeneca also led to the production of tomatoes having delayed ripening, but by using a different gene-silencing mechanism. The starting point was again the identification of the gene expressing the enzyme PG, but its action was blocked using sense gene suppression.[3] The mechanisms are still unclear, but the integration of a sense gene construct having a coding sequence identical (homologous) or similar to the target PG gene resulted in the inactivation of the PG gene. This prevented the breakdown of cell walls but, like the antisense technique, left other changes associated with ripening, such as colour and flavour development, to continue as normal. These tomatoes were not grown in Britain, as they are better suited to warmer climates, and were predominantly grown for use in processed food.

Tomato purée from Zeneca's tomatoes first appeared in Sainsbury and Safeway supermarkets on 5 February 1996. Purée from genetically modified tomatoes has been clearly labelled by voluntary agreements by both supermarket chains and, according to the Food and Drink Federation, sales have been good.[6] In February 1997, Zeneca applied to have its modified tomatoes sold as whole tomatoes or as diced tomatoes in cans,[7] but, as with the Calgene tomatoes, it has to wait for marketing approval through the EU Directive on Genetically Modified Organisms, which may take up to two years.

It is estimated that up to half the fruit and vegetables grown commercially are lost to spoilage. Besides tomatoes, a number of research groups are now aiming to produce a range of other slow-ripening fruits and vegetables with a longer storage life to reduce wastage. Ethylene is a key plant hormone in many physiological and developmental processes, including the ripening process and the shedding of leaves and flowers. Ethylene is produced naturally to control the ripening of climacteric fruits, that is those fruits that change their pattern of respiratory gases during ripening, such as banana, tomato, apple, pear, mango and melon. Bananas, which are picked and transported unripe, are ripened by the application of ethylene gas. In climacteric fruits, inhibition of ethylene synthesis could inhibit ripening. Non-climacteric fruits, such as orange, lemon and strawberry, do not alter their respiration during ripening and, therefore, are not amenable to ethylene manipulation.

Antisense DNA can be used to inhibit ethylene synthesis in climacteric fruit crops.[8] The ethylene synthesis metabolic pathway is under the control of two enzymes, called ACC synthase and ACC oxidase. The process, once started, is autocatalytic, that is one of the products of the process acts to promote the process. Melons (*Cucurbita melo*) of the cantaloupe Charentais type, chosen for their good eating quality but poor storage capability, were produced with an antisense gene for ACC oxidase, thereby blocking the last stage of the ethylene biosynthesis pathway and inhibiting ripening.[9] These melons remained on the plants because they did not develop the characteristic abscission layer of cells that causes fruit to separate from the plant, although the flesh developed as normal. After ten days under storage conditions, the antisense cantaloupes had firm green rinds, while control fruit of the same age had shrivelled yellow rinds, with fungal infection and squashy soft flesh. The antisense effect can be reversed by treating fruits with ethylene. Fruits can therefore be picked unripe, stored for long periods and ripened as required by ethylene treatment. The gene-silencing techniques developed at the University of Nottingham can also be used to delay ripening in climacteric fruits by switching off the genes that express ACC synthase and ACC oxidase. Research in this area may eventually lead to improvements in the quality, length of storage life, appearance and nutritional value of many climacteric fruit crops.

The application of genetic engineering promises great benefits for the post-harvest storage of perishable crops, with reductions in spoilage and waste. This could bring benefits to developing countries, enabling more fresh fruit to reach consumers. However, against the considerable advantages, it should be noted that as food ages it may decrease in nutritional value, while consumers are prevented from determining the true age of the food based on their past experience. This is good for food producers and supermarkets, as less wastage leads to more profits, but not necessarily good for the consumer. However, independent assessment of the nutritional information supplied by companies is now done as part of the marketing approval process.[5]

The composition of potatoes has also been changed by genetic engineering, but for different reasons than for tomatoes. A gene from a starch-producing strain of the bacterium *Escherichia coli* was integrated into potatoes to increase their starch content. This gene expressed an enzyme called ADP-glucose pyrophosporylase, resulting in tubers with up to 20 per cent more starch than unmodified potatoes.[10] This increase in the solid content has no effect if potatoes are baked or boiled, but is advantageous when they are fried during the making of fries or crisps. Water is replaced by oil during the frying process. The high-starch potatoes contain less water and so absorb less oil. French fries usually contain around 36 per cent oil, whereas fries made from transgenic potatoes contain 30 per cent oil. The resulting fries are more nutritious and, with less oil, healthier, than regular potatoes. The transgenic potatoes also need less energy to cook, as the energy used in frying goes towards removing water. A team at Monsanto has commercially developed these potatoes,[11] and McDonalds are starting to use them for their French fries. The potatoes are quick-frying, which results in a faster turnover.

Transgenic crops have also been engineered with in-built sweetness. Genes expressing proteins with many times the sweetness of sucrose have been expressed in a number of crops. Thaumatin, a protein naturally produced in the West African plant katemfe (*Thaumatacoccus danielli*), has been expressed in potatoes; monellin, produced naturally in serendipity berries (*Dioscoreophyllum cumminisii*), has been expressed in tomato and lettuce.[12] Levels of expression are low, around 1 per cent of total protein in the case of monellin, but both these proteins are around three thousand times the sweetness of sucrose. These expression levels are therefore sufficient for the sweetness to be detectable to human taste. Calgene have, meanwhile, identified a gene with which they hope to produce high-sweetness strawberries (*Fragaria chiloensis*). The transgenic expression of sweetness will lead to a range of designer tastes in agricultural products in tomorrow's supermarkets. Transgenic fruits may also soon be produced without unwanted pips. A gene called SDLS-2, found in a range of plants, is responsible for killing unwanted cells during the plant's development. The gene has been linked to a promoter that makes it destroy seeds. Researchers in Australia and Japan have genetically modified tobacco to destroy its own seeds and the technique is being adapted to produce pipless oranges.[13]

Oilseed composition

Oilseed composition is another area where modification of food for health reasons is being investigated. Medical research has stressed the benefits of less fat in the diet and a shift away from saturated hard fats to polyunsaturated soft fats. The former are associated with dairy products and the latter with vegetable oils. Enzymes exist in plants that convert one type of fat to the other. The genes that code for production of these enzymes have been identified, and transferred into oilseed crop plants. Although both hard and soft fat types occur in vegetable oils, a shift in the balance between the fat types present in vegetable oils can be obtained by genetic engineering. Calgene is developing vegetable oils from transgenic canola, a close relative of oilseed rape, containing a greater proportion of polyunsaturated fatty acids, and a greater proportion of short- to medium-chain fatty acids, than traditional vegetable oils.[14] These will be sold as food ingredients for the medical and health food markets.

Oilseed rape and canola have been modified to produce a range of specialist oils. Calgene has patents on a wide range of transgenic canola varieties. Calgene's high-laurate canola was the first genetically modified oil to be sold commercially, when it was approved for the Canadian market in 1996. Calgene now markets a range of high-laurate oils produced from transgenic canola under the Laurical brand-name. Laurate is not normally present in canola or any other non-tropical crop. It is used in food additives, including confectionery coatings, coffee whiteners, reduced fat-cream and dairy cream cheese, and in detergents, and traditionally comes from coconut and palm kernel oil suppliers in the Third World (see Chapter 14). The canola was modified using a gene from the Californian bay laurel tree (*Umbellularia california*) which, on expression, produces an enzyme called thioesterase. This enzyme acts on the existing biochemical pathways to produce the novel lauric fatty acids.[15]

Calgene is also developing high-stearate canola oil. This will create a potential alternative to hydrogenated oils in margarines, food shortenings and confectionery products. Recent medical evidence has suggested that substances called *trans* fatty acids, formed during the process of hydrogenation, which turns liquid vegetable oils into solids for margarines and other food products, may be harmful to health. High-stearate vegetable oils might eliminate the need for hydrogenation. Canola oils that mimic castor oil and other specialist oils are also under development.

Protein content

Animal protein contains all 20 essential amino acids. Plant seed proteins are deficient in some of these essential amino acids. Different crop plants are deficient in different amino acids. Maize storage proteins are low in lysine and tryptophan, while legumes are deficient in the sulphur-containing amino acids cysteine and methionine – soybeans, for example, lack methionine. At the present time, to prevent dietary deficiency in a vegetarian diet, a combination of plants or plant products is consumed. Combinations of food plants, such as rice and beans (or baked beans on toast), are eaten to supply all the essential amino acids. However, genetic manipulation could be used to supply all the essential amino acids in one transgenic food plant.

Seed storage proteins are the main candidates for genetic modification to enhance the nutritional value of plants through the manipulation of protein. The genes coding for a number of plant proteins have been identified and cloned. A gene for a wheat storage protein called glutenin has been expressed in tobacco, while the zein gene, present in cereals, can be expressed in a range of broad-leaved crops if a suitable promoter gene is available.[16] The phaseolin gene from legumes has also been transferred to other species. Transgenic soybeans and canola, engineered with a gene from brazil nut (*Bertholletin excelsa*), have produced methionine-rich soybean seeds and methionine-enriched oil, respectively.[17]

Viral resistance

Viruses consist of a core of nucleic acid surrounded by a protein coat. They need to infect a cell of another organism in order to reproduce. In plants, infection can cause a range of diseases, often involving the yellowing and blistering of leaves. Viral diseases cause economic damage in most of the major agricultural crops. No chemical viricides are available that also leave crops unharmed, although insecticides can be used to control the insects that spread viruses. Genetic modification could therefore make a significant contribution in this area. Some plants have natural resistance to viral infections, while in others a mild infection can result in a natural inoculation against subsequent challenges. The 'induced resistance' principle employed in the development of most virus-resistant transgenic plants is similar to that used to produce vaccines in animals. Genes coding for proteins found in viral coats, or other viral gene sequences, are integrated into the plant genome.[18] These coat proteins 'prime' the plant to fight the real virus when it is encountered. The reaction is specific to a particular group of viruses and plants remain susceptible to other viruses.

Monsanto has developed tobacco and tomato plants carrying genes for coat proteins of tobacco mosaic virus, which confer partial resistance to tobacco and tomato mosaic viruses. In this case, the presence of the coat proteins in the transgenic plant interfered with an early stage in the replication of virus particles in the plant. Viral coat protein genes have also been engineered into potato, against potato leaf roll luteovirus and potato viruses X and Y, and papaya, against papaya ringspot virus.[19] Monsanto is involved in work to produce a sweet potato resistant to feathery mottle virus, and Agrigenetics Advanced Science, Pioneer Hi-Bred, Upjohn and other companies have field-tested a range of other virus-resistant transgenic crops, including alfalfa, cucumber, cantaloupe and squash. In August 1997, Monsanto applied for a permit to market transgenic potatoes carrying a gene from the potato leaf roll virus.

Viral resistance can also be obtained by integrating other viral RNA genes into plant genomes. Cucumber mosaic virus (CMV) is one of the most widely occurring plant viruses, infecting more than eight hundred species, including a wide range of horticultural crops. In addition to an RNA genome, CMV strains also have RNA structures called satellite RNA. Satellite RNA does not code for protein and has little similarity to the main viral genome, but is capable of modifying the infectivity of the virus.[20] Satellite RNA can attenuate CMV strains, resulting in a sharp decline in the virus and an almost complete lack of symptoms. Plants have been engineered to express satellite RNA genes, resulting in a high degree of tolerance to CMV infection.

A number of unique ecological risks are associated with virus-resistant transgenic crops, however, which may limit their deployment in field situations (see Chapter 7).

Fungal resistance

Fungi cause yield losses in a number of major crops. The usual treatment is by spraying fungicides. Antifungal proteins, however, occur naturally in tobacco and other plant species. The genes expressing these proteins have been used to produce transgenic plants that are resistant to a number of fungal diseases.

The cell walls of fungi contain chitin (unlike plant cell walls, which contain the related chemical cellulose). The enzyme chitinase, the major antifungal protein, acts to break down the fungal cell wall. Other antifungal proteins may trigger additional defensive reactions against fungi. Resistance to the fungal pathogen *Rhizoctonia solani*, responsible for damping off and seedling blight, has been, for example, obtained in tomatoes by integrating a gene expressing a chitinase isolated from French bean. A chitinase has also been expressed in tobacco, the model crop for this research, against the brown spot fungus, *Alteria longipes*.[19]

Another approach is to programme infected plants to commit suicide. Potatoes, genetically engineered so that their cells die if they are infected with fungal disease, will act to prevent disease spreading through a field. Scientists from the Max-Planck Plant Breeding Institute in Cologne, Germany, inserted a gene that coded for barnase, from the bacterium *Bacillus amyloliquefaciens*, into potato.[21] This enzyme destroys ribonucleic acids, including mRNA, and therefore stops all protein synthesis. The gene was attached to a potato promoter gene before insertion, with the result that if the cell is infected the promoter switches the barnase gene on, effectively killing the plant. Field trials were conducted in 1996 with the potato variety Bintje, which under field conditions is sterile, therefore reducing the risks of genes being transferred to wild relatives. If successful, the approach could lead to large reductions in fungicide use against potato blight and other diseases.

Nematode resistance

The nematodes comprise a surprisingly large number of free-living and parasitic roundworms. Plant-eating nematodes live in the soil and feed on roots, causing crop losses of around US$100 billion annually. Two groups of nematodes, the root-knot worm and the cyst nematode or eelworm, account for most of this crop damage. Traditional nematode control includes soil fumigation before planting and the use of chemicals, which are both toxic and expensive. One of the most effective fumigants, methyl bromide, is being phased out or banned in many countries because it damages the ozone layer.[22]

A group at the Dutch Centre for Plant Breeding and Reproduction Research in Wageningen has identified a plant gene that confers nematode resistance. The gene was isolated from a wild beet known to be resistant to nematodes that damage commercial sugar beet. However, crossing this beet with commercial varieties using traditional plant-breeding methods only resulted in weak plants. Dutch and Danish companies are now producing transgenic sugar beet varieties, with the wild beet gene, which have nematode resistance.[22]

A related approach is being developed by a group at Leeds University in England, whereby genes expressed in one part of a plant, which confer nematode resistance, are transferred to another part of the plant where the genes are not normally expressed.[23] For example, nematodes do not eat rice grain because of the presence of protease inhibitors, which prevent the worms digesting the available protein. Transferring the genes that express these protease inhibitors to rice roots should confer enhanced levels of nematode resistance. Transgenic rice varieties of this type could have a wide application in the developing world, as they would be resistant to many of the different nematode species attacking crops. The group at Leeds, working in collaboration with Advanced Technologies (Cambridge), have also developed potatoes resistant to the root-knot nematode, by transferring genes expressing protease inhibitors into the roots. These genes will not be expressed in the potato tubers that are eaten.[23] Advanced Technologies have patented this NemaGene™ nematode resistance technology, for use against root-knot nematode and cyst nematode in a range of crops, including potato, tomato and sugar beet. This approach may bring great benefits to Third World countries such as Bolivia, where potatoes form an important part of the diet, and where nematicide treatments are prohibitively expensive.

Photosynthesis and nitrogen fixation

Many significant crop improvements have already been made using genetic engineering. There are many more modifications at the development stage that may have great significance in the future. The crop transformations that could potentially make the most difference to the world food supply would be, first, improving the efficiency of photosynthesis, and, second, extending the ability of plants to fix nitrogen.[14]

Photosynthesis is the chemical process green plants use to synthesize organic compounds, from carbon dioxide and water, using energy from sunlight harnessed by chlorophyll molecules. Each of the many stages in this process involves a specific enzyme. As seen throughout this book, enzymes can potentially be modified by genetic manipulation. There is much potential for improvement, as photosynthesis is known to be less than optimally efficient. The photosynthetic pathway occurs in specialized structures called chloroplasts. Water is first split into its hydrogen and oxygen components, then the hydrogen is attached to carbon dioxide captured from the atmosphere to form organic molecules, the first of which is glucose. Glucose is passed into complex biochemical pathways, to be built up into amino acids, starches, fats and cellulose.

A key enzyme, which brings carbon dioxide into the metabolic cycle, is ribulose bisphosphate carboxylase, usually abbreviated to rubisco. This enzyme can account for up to 50 per cent of the protein within a green leaf. However, rubisco, in addition to binding ribulose bisphosphate to carbon dioxide, also binds it to oxygen, thus breaking it down to form carbon dioxide, in a process called photorespiration. A research programme at Rothamsted Experimental Station, in England, was established to redesign the rubisco molecule by altering the coding sequence of the gene for rubisco, to reduce or eliminate the oxidation reaction of the molecule, while leaving its carbon dioxide capturing function unimpaired.[24] This would make photosynthesis more efficient.

There are two different types of photosynthetic pathway in plants. Most crops of temperate regions combine carbon dioxide with rubisco, as described above, to form two molecules of the 3-carbon phosphoglyceric acid (PGA). These are called C3 plants. Plants of this type include soybean, wheat, oats and potato. However, many tropical crops, including tropical grasses, first attach carbon dioxide to a 4-carbon molecule called oxaloacetate, and these plants are called C4 plants. Plants of this type include maize and rice. The C4 pathway is more elaborate and more efficient than the C3 pathway, so given the same amount of light, C4 plants will fix more carbon for photosynthesis. Research programmes have been established to see whether transferring C4 pathway photosynthesis to temperate crops would bring benefits.

Nitrogen is the most abundant gas in the atmosphere, although it is not directly available to most organisms from the air. Nitrogen is essential to the growth of all living organisms. Plants pick up nitrogen in the form of nitrates from the soil, although some groups of plants obtain nitrogen more efficiently through a symbiotic association with bacteria. In leguminous plants, nitrogen fixation occurs in root nodules with the assistance of *Rhizobium* soil bacteria. The bacteria have genes that express nitrogenase enzymes, which convert atmospheric nitrogen to ammonia, and hence to amino acids. The root nodules provide the bacteria with a carbohydrate supply and a site that excludes oxygen, a gas that terminates the nitrogen fixation process. Modifications could theoretically be made to *Rhizobium* bacteria to make them more efficient, while genes for nitrogenase enzymes could be transferred to leguminous plants to remove the requirement for bacteria.

In experiments with millet (*Panicum miliaceum*), growth was up to 17 per cent faster when the soil around the root was inoculated with the nitrogen-fixing bacterium *Azospirillum lipoferum*. Nitrogen fixation is, however, a complex process, with 17 bacterial genes involved. Nitrogen fixation is also an energy-consuming process, and transgenic nitrogen-fixing cereals may yield less than cereals fertilized artificially. They could, however, prove useful in places where fertilizer use is impractical or undesirable for environmental reasons.[25]

Tolerance to high salinity and other poor soil conditions

Tolerance to a range of environmental conditions can be selected for in the tissue culture stage of transgenic plant production. Tissue culture usually results in the production of genetically identical plants (clones), but some individuals will alter their genetic make-up. This is referred to as somaclonal variation. This can be exploited, by using a large number of clonal cells and applying a particular environmental stress factor. For example, high salt concentration can be applied in the tissue culture nutritional solution. The surviving cells can be grown as salt-tolerant plants. These plants could also contain foreign genes coding for other desirable characteristics.

The Tissue Culture for Crops Project at Colorado State University, USA, is active in this research area, particularly in developing rice tolerant to saline conditions. Conventional plant breeding cannot be used to breed for two distinct characters simultaneously. However, genetic manipulation techniques allow salt tolerance and high-yield characters to be integrated into the same transgenic plant. One method of achieving this is by protoplast fusion. The protoplasts (cells stripped of their cell walls using enzymes) can be fused with the aid of chemical treatments. Protoplasts of a wild rice strain, found in salty mangrove swamps in Bangladesh, have been successfully fused with protoplasts from a food rice using this method. The mangrove rice strain has microscopic structures called salt hairs on its leaves, in which excess sodium chloride is accumulated.

Transgenic salt-tolerant plants are also being produced, by incorporating a gene from a yeast that survives in salty environments. The gene expresses a protein that causes sodium to be pumped out of cells, thereby reducing the damage caused by high sodium levels in the soil. Salt-tolerant tomato, melon and barley varieties are at the developmental stage.[26]

Soil can contain high levels of metals or other contaminants. Transgenic crops could be produced to tolerate these conditions – tobacco tolerant of high cadmium levels, for example, has been produced by integrating a gene from a mouse expressing metallothionein-binding protein.[27] Plants are also used to clean up contaminated land, a process called bioremediation. Transgenic crops may come to play a role in this area in the future.

Tolerance to drought conditions

Drought-tolerant plants could potentially be engineered with roots that penetrate deeper through dried-out soils, have thicker cuticle to minimize water loss, or the ability to make adjustments to the salt content in their cells. The first drought-resistant transgenic plants were produced using a gene from baker's yeast that expressed trehalose. This substance enables yeast to survive in a dried state. Tobacco plants, modified with this gene, withstood desiccation that left control plants dying.[28]

Transgenic drought-tolerant plants would have extended growing seasons and extended ranges in places where water is a limited resource. Therefore, these transgenic plants could be beneficial not just under desertification conditions, but in many agricultural situations where irrigation is needed. Drought-tolerant varieties could conserve valuable water resources and, given current estimates of global warming rates, may become important in many regions in the future, just to keep crops growing within their present range.

Tolerance to frost: ice minus bacteria and antifreeze proteins

Frost damage accounts for around US$4 billion worth of crop losses annually. The damage is caused by the formation of ice crystals inside cells, which causes structural damage and renders the plant tissue soft when it thaws. Frost damage in the field also leads to rotting and wastage of fruit and vegetables. Ice formation is initiated at regularly shaped surfaces, and with plants ice tends to form on the coat proteins of bacteria living on their surface. An early application of genetic engineering to crop production was the ice minus bacteria, one of the first releases of a genetically modified organism into an agricultural ecosystem. A mutant of the bacterium *Pseudomonas syringae* was isolated, whose coat proteins had been altered so that it no longer provided a regular surface for ice formation. The gene involved was identified and called ice, and the mutant lacking the gene was called ice minus. The gene expressing the ice-formation protein was deleted from *P. syringae*, by Stephen Lindow and colleagues at the University of California in Berkeley, to form genetically engineered ice minus bacteria.[29]

It was therefore not the crops themselves that were altered, but the associated bacteria. Suspensions of ice minus were sprayed onto crops, coating the plants. The modified bacteria competed with and displaced the resident bacteria on the plants' roots. Strawberries are particularly prone to frosts, and were the crop on which ice minus was initially tested. The crop became frost resistant. Potatoes and tomatoes were also used in early successful trials of ice minus. Advanced Genetics Systems carried out the trials on strawberries in California, while ice minus was marketed in the USA by Monsanto.

Crop plants themselves may soon be genetically modified to resist frost damage. Frost-resistant fruits and vegetables could be stored at sub-zero temperatures, for longer periods than susceptible produce, without losing their texture or flavour. Therefore, engineering frost resistance into fruits that are poor freezers, such as strawberries and tomatoes, could prove economically beneficial to food retailers. An additional advantage in the field will be that frost-resistant crop varieties could enjoy a longer growing season and have a larger geographic range.

Two main approaches have been tried in order to modify crops to resist freezing: altering their fat composition and integrating genes expressing antifreeze proteins. The former technique involves the same approach as used to change the composition of fats in oilseeds for nutritional and health reasons. The composition of lipid molecules in cell membranes changes according to temperature. Plants surviving in low temperatures shift the balance of their fat composition towards unsaturated lipids, which are more fluid at lower temperatures. This maintains the integrity of cell membranes under cold conditions, reducing frost damage. Genes coding for enzymes that convert saturated to unsaturated lipids have been identified in many plants.[30] Cold-resistant plants are more effective at switching these genes on, and crops are likely to be engineered with such genes in the future.

A gene expressing an antifreeze protein has been identified from the winter flounder (*Pseudopleuronectes americanus*), an Arctic fish that can survive temperatures that would freeze most other fish. The protein binds to the water–ice interface and prevents ice crystals from forming. The gene has been used to produce transgenic frost-resistant tomato and tobacco plants.[31]

Drugs and vaccines

One important development in transgenic crop production has been the experimental integration of genes expressing therapeutic drugs or vaccines. Crops may one day become as commercially important as the cows and sheep that produce human proteins in their milk. Crops have several advantages over animals as 'bio-reactors'. Large quantities of plant material can be easily produced, fewer ethical concerns are involved and crops such as modified bananas could provide an easy source of medicinal drugs, particularly in the developing world. In 1997 Applied Phytologies, a company based in Davis, California, grew a crop of transgenic rice that expressed alpha-1-antitrypsin in its grain.[32] This human protein has also been expressed in transgenic sheep's milk (see Chapter 3). Crops that are eaten raw are preferable for vaccine production, as cooking can denature many therapeutic products. Bananas have been genetically modified to carry hepatitis B vaccine, and it was estimated that ten hectares would produce enough to vaccinate all the children in Mexico.[33]

An important technical advance was the production of a chimaeric virus particle, a combination of a plant virus and a gene from a human or animal virus. This modified virus can be grown in whatever crop the original plant virus infects, to produce an effective vaccine.[34] A range of vaccines are now being produced in bananas, cowpeas, and other crops.

Engineering cotton: blue jeans and plastic plants

Transgenic crops will soon be used to produce raw materials for industry. Cotton is one of the most successful crops to have been genetically modified. In 1997, almost a quarter of the total US cotton crop was grown from transgenic seed. This cotton was resistant to major insect pests of cotton, principally *Heliothis* species, or resistant to the herbicides bromoxynil or glyphosate. Clothing, including T-shirts, made from genetically modified cotton is already exported around the world. Genes are now being developed that modify the properties of cotton.

Monsanto's blue gene project aims to develop cotton plants containing foreign genes that express blue pigment, for the blue jean market. In 1997, the company had already obtained blue lint from cotton plants. Coloured cotton fibre would reduce the need for dying and provide unique colour-fastness. Meanwhile, Agracetus has developed transgenic cotton with fibres containing a polyester-like compound, while its parent company Monsanto has patented a number of genes that produce plastic materials in transgenic crops. Genes that synthesize the biodegradable plastic polyhydroxybutyrate (PBH), for example, have been expressed in plants.[35] The potential for producing materials for industrial use is enormous. Genetically modified crops are therefore starting to make major contributions in a number of areas, in addition to food production.

Notes

1. Calgene was founded in 1980 to develop genetic engineering for agricultural applications. The company has concentrated on three areas: fresh produce, in particular tomatoes and strawberries; speciality oils from transgenic canola; and cotton.
2. Pear et al., 1993.
3. Grierson et al., 1987.
4. In March 1996, Calgene and Monsanto entered into a transaction under which Monsanto contributed Gargiulo Inc., US$30 million and certain oils and

produce-related technology to Calgene in exchange for a 49.9 per cent equity interest in Calgene. Gargiulo is a grower, packer, marketer and distributor of tomatoes and strawberries. In November 1996, Calgene and Monsanto closed a transaction whereby Monsanto purchased enough shares to give it a 54.6 per cent equity interest. In January 1997, Monsanto gained total control of Calgene. http://www.calgene.com/

5. Advisory Committee on Novel Foods and Processes (ACNFP), Press Release, 'Report on Genetically Modified Tomato to be Eaten Fresh'. ACNFP, MAFF, London, UK, February 1996.

6. Food and Drink Federation, London, UK, 1996, *Modern Biotechnology: Towards a Greater Understanding.*

7. *Daily Mail*, 3 March 1997, p. 9.

8. Lycett and Grierson, 1990.

9. Ayub et al., 1996.

10. Barry et al., 1992.

11. *Science* 257: 1480, 11 September 1992.

12. Penarrubia et al., 1992.

13. *New Scientist*, 29 March 1997, p. 11.

14. Voelker et al., 1992.

15. Voelker et al., 1996.

16. Hoffmann et al., 1987; Roberts et al., 1989.

17. Altenbach et al., 1992; Townsend et al., 1992.

18. Abel et al., 1986; Greene and Allison, 1994.

19. Dale et al., 1993.

20. Tepfer and Jacquemond, 1996.

21. *New Scientist*, 11 May 1996, p. 20.

22. *New Scientist*, 31 May 1997, p. 12.

23. *Guardian, Online*, 13 March 1997, p. 10.

24. Tudge, 1988.

25. Tudge, 1993.

26. *New Scientist*, 26 July 1997, p. 16.

27. Maiti et al., 1989.

28. *Guardian, Online*, 21 August 1997, p. 7. Trehalose may also protect plants from freezing damage.

29. Lindow and Panopoulos, 1988.

30. For example, a gene expressing glycerol-3-phosphate acyl transferase from *Arabidopsis thaliana*. Murata et al., 1992.

31. Hightower et al., 1991.

32. *New Scientist*, 12 July 1997, p. 17.

33. *New Scientist*, 21 September 1996, p. 6.

34. *Observer*, 25 May 1997, p. 12.

35. Poirier et al., 1992.

7. Ecological risks

On their release into the environment, transgenic organisms can present a number of potential ecological risks. Estimating these risks, however, is problematic as regulatory organizations have little previous experience of modified organisms in the environment. Transgenes are inheritable and will appear in the genomes of the offspring of either genetically modified organisms or organisms that acquire the transgene by whatever mechanism. Therefore, once spread to the wider environment, transgenes may be impossible to eradicate. This chapter examines the ecological risks posed by a range of transgenic organisms.

Micro-organisms present particular challenges, because of their rapid reproductive rate, their willingness to exchange genetic material, and the difficulty of detecting them in the environment. The major ecological concerns with genetically engineered crops are: a) that they may, by gaining in vigour or invasiveness, become weeds of agricultural or natural habitats, and b) that genes may be transferred from them to wild relatives, whose hybrid offspring become detrimental in some way to the existing flora or fauna. Transgenic fish and animals present different sets of ecological risks.

Risk assessment

Experimental trials with transgenic organisms are usually conducted under strict regulations to minimize the potential spread of genetic material. To comply with federal US regulations, for example, genetically modified plants have to be transported in sealed containers, and plots of experimental transgenic crops are surrounded with moats, fences and vegetation-free areas, while mature plants are stripped of pollen-bearing and other reproductive parts. Even given these regulations, however, no field trial can be said to be 100 per cent secure. This was illustrated when flooding struck the American Midwest in July 1993 and an entire field of experimental insect-resistant maize was swept away in Iowa.[1] No material was ever recovered and the plants were probably buried under several feet of river mud. A spokesman for Pioneer Hi-Bred International, the company involved, explained that at the time of the flood the plants were too young to transfer genetic material to other plants. However, a more unfortunately timed natural disaster might have resulted in the spread of genetic material. Once released accidentally into the environment, plant material may prove very difficult to recover.

The Iowa incident illustrates the difficulty of predicting the ecological risks of releasing transgenic crops to the environment. It is probable that some transgenic organisms released to the environment will become established despite the safeguards. In most cases they are likely to pose no threat to agricultural or natural habitats. However, it is little understood why one species becomes a weed, while another closely related species does not. A report by the British Royal Commission on Environmental Pollution,[2] in an attempt to quantify risks, reasoned that the ability to predict the outcome of a release was likely to be greater if the transgenic organism released was a modified version of an organism common in the locality of the release site. Predictability would also be increased if the modification was limited in scope, the properties of the new genetic material were well understood and the quantities released were not excessive. Most foreign genes already exist in nature, albeit in other organisms, and should act more predictably than genes with altered coding sequences.

The Royal Commission report recommended the HAZOP (HAZard and OPerability) technique as a structured and systematic approach to the identification of risks. The technique was developed to expose hazards in chemical factories and involves teams of experts identifying unplanned events that might occur during day-to-day operations. This technique could draw attention to previously unforeseen ways in which transgenic crops could pose ecological risks, but cannot provide factual answers about the probability of events occurring.[2]

The creation of artificial environments or microcosms, small-scale trials in greenhouses, and caged areas in the field can provide useful information on the stability of modified organisms, their gene transfer potential and the effects of environmental parameters on gene expression. However, carefully monitored small-scale experiments are very different from large-scale commercial releases of transgenic crops. The commercial situation involves a much greater number of seeds, and more opportunities for interaction with wild relatives. A step-by-step approach, from laboratory to greenhouse to experimental field trial to monitored large-scale release, is desirable in assessing risk. Different organizations will be responsible for the scrutiny of experiments at different stages (see Chapter 11) and, therefore, close cooperative links between regulatory bodies are desirable.

There is a great deal of unpredictability concerning the ecological impacts of transgenic plants. Therefore, studies on ecological parameters themselves risk 'genetic pollution' of the environment.[3] If only 'safe' plants were released in experiments to quantify risks, then the results would suggest that the technology was trouble-free.[4] However, too many unknowns exist for theoretical work to be useful, so risk assessment based on field experiments has proved to be the way forward, using marker genes to track behaviour of model genetically modified organisms in the environment. Commercial plantings are not designed to yield ecological insights, so there is a need for large-scale experiments with model transgenic varieties to study invasiveness and gene spread to wild relatives.

Risks posed by transgenic micro-organisms

Micro-organisms pose particular ecological risks because of their short generation time and high mutation rates, and their ability to pass genetic information between themselves, in a process called conjugation. Millions of offspring, containing copies of a transgene, could be produced within days, or even within hours. The difficulty of their detection in the environment also means that there is a particularly high level of uncertainty associated with risk assessment involving genetically engineered micro-organisms. Detection is usually by sampling and incubating soil samples on growing media, but certain micro-organisms may escape detection. There is also a basic uncertainty about how micro-organisms live, as their natural ecology is often not well understood. The long-term monitoring of a genetically modified micro-organism, however, requires a good knowledge of the microbe's ecology.[5]

Different soil types have been shown to influence the behaviour of genetically engineered micro-organisms in the environment.[5] Large amounts of DNA are naturally added to the environment as a result of excretion, death and decay. This DNA is called free DNA and is usually rapidly degraded. DNA from transgenic organisms, however, may persist in the soil under certain conditions. For example, in clay soils DNA adheres to small soil particles where it becomes more resistant to degradation. This DNA could be taken up by soil bacteria.

The ice minus bacteria, *Pseudomonas syringae*, engineered to lack an ice nucleating protein, competes with non-modified micro-organisms to produce frost resistance in strawberries and potatoes onto which it is sprayed (see Chapter 6).[6] The original applications by Advanced Genetics Sciences for field release were made in 1984. Approval was given in 1986 and the first experimental release was on a small field plot at a University of California research station in April 1987. The site was vandalized by protesters the following month. It has been claimed that the company also performed an illegal open-air experiment with ice minus bacteria on the roof of their laboratories prior to any release being authorized, which would have given the micro-organism ample opportunity to disperse.[7] The vandalization of experiments involving genetically modified organisms has occurred on a number of occasions in the USA and Europe, which increases the risk of transgenes being dispersed into the environment.

A major concern with the use of ice minus bacteria was that the genetically modified microbes might persist in the environment. Extensive monitoring during an 18-month trial at Clemson University's experimental research station in North Carolina showed that the bacteria remained close to plants on which they were sprayed. Subsequent studies have not detected *P. syringae* in areas surrounding the spray zones.[6] Ice minus bacteria persisted for about a week after spraying in the soil on the experimental site.

In order to study the persistence of bacteria in the soil Monsanto integrated a gene from *Escherichia coli* that acted to split a chemical analogue of lactose called 'X-gal', into another species (*Pseudomonas aureofaciens*).[8] The engineered bacteria producing a bright blue-coloured chemical when the soil was treated with a selective medium containing lactose as its only energy source. The technique is very sensitive, allowing a single bacterium in a gramme of soil to be detected. Critics argued, however, that if the bacteria containing the marker gene had entered run-off water, its ability to break down the milk sugar lactose could have threatened any local dairy industry.

The Environmental Protection Agency (EPA) also monitored the dispersion of ice minus bacteria in the air, in a small-plot study in California.[6] They found no dispersion to neighbouring vegetation. Spraying during still conditions is likely to reduce spray drift and the spread of bacteria in the environment. In their thorough study of the possible ecological risks of genetically modified organisms, the Royal Commission on Environmental Pollution[2] postulated that if ice minus bacteria were to become widespread in the atmosphere, they might lead to changes in local climate, preventing the formation of rain droplets. The risks were estimated as negligible, but illustrate the type of risks that need to be considered with genetically engineered micro-organisms.

Micro-organisms modified with genes that cause them to resist degradation, either by design or as a side-effect of another characteristic, could pose a greater ecological risk because they are present in the environment for longer. Modified organisms are, however, more likely to be debilitated in some way before their release to ensure they do not persist in the environment. This can be done by making some change in the genes controlling metabolic events, to make the organisms less competitive in the environment, or by incorporating suicide genes, which disable the organisms after they have fulfilled their desired role. Genetically engineered baculovirus were 'crippled', during early experimental releases against insect pests, by removing their coat proteins. This made the viruses less stable and less likely to survive in the environment for long periods. This approach may prove successful for some micro-organisms, but the disabled baculovirus proved less effective at killing insects. In this case an inability to persist in the environment would reduce the commercial benefit of the baculovirus (see Chapter 5).

The baculovirus study also highlighted the need to screen native species in nearby natural habitats for susceptibility to the engineered control agent. The baculovirus was engineered to be a more effective killer of the caterpillars of moth pests on cabbage and other vegetable crops. A number of native moth species were found to be susceptible to the baculovirus, although only at high doses.[9]

Risks posed by virus-resistant crops

Unique ecological risks have been associated with virus-resistant transgenic crop plants. These transgenic crops have sequences of viral nucleic acid integrated into the plant genome (see Chapter 6). Resistance to cucumber mosaic virus, for example, was obtained in crops by transferring a gene sequence from satellite RNA, a structure in the virus that can reduce viral symptoms. However, under some circumstances satellite RNA can exacerbate rather than reduce viral symptoms. This was observed in Italy in 1996, when a naturally mutated satellite RNA sequence led to an epidemic of lethal tomato necrosis, which resulted in severe crop losses. Deleterious forms of satellite RNA were thought to be rare events in nature, and therefore unlikely to occur in transgenic plants. However, recent work has shown that deleterious satellite RNA can arise by mutation more commonly than previously thought, and that it has a selective advantage over the parent satellite RNA.[10] It would therefore be hard to eliminate such mutations if they were to occur within sequences contained in transgenic plants, leaving crops more vulnerable to virus attack and risking the spread of virus susceptibility to other plants.

Viruses have been shown to pick up genes from transgenic crops. In laboratory experiments, viruses from which genes for particular characters had been removed reacquired those genes from transgenic plants that had them inserted into their genomes. This transfer of viral genetic material means that other plants or viruses could pick up the transgenes. In one Canadian study, plants were infected with cucumber mosaic virus that lacked a gene for a protein, but the viruses were able to acquire this gene if it was engineered into the plant's genome from another virus.[11] Wild viruses can therefore take transgenes consisting of viral genetic material from the genomes of crop varieties designed for viral resistance. A growing concern that new hybrid viruses could be produced led the United States Department of Agriculture (USDA) in August 1997 to outline proposed restrictions on transgenic crops engineered with genetic material from viruses. These restrictions included limits on the length of genetic sequences integrated into plants and the banning of the use of particular genes.[11]

Virus-resistant crops will contain viral genes in all their cells for the lifetime of the plants, and given the ability of viruses to acquire, recombine and swap genetic material, the deployment of large areas of these transgenic crops may create the ideal conditions for new disease-causing viruses to evolve.

Risk of invasion and adverse effects on other organisms

Transgenic organisms might become more vigorous or invasive and themselves become weeds or pests. Many of the world's weed and pest problems arose from exotic introductions – organisms that were transferred from their native habitats to ones in which they were not normally found. These exotic introductions provide a model for assessing a worst-case scenario for the potential effects of a modified organism that changes to become more invasive. Exotic introductions can be accidental or can be the result of deliberate releases that have resulted in unforeseen ecological effects. They can achieve levels of population growth impossible in their native habitats, due to increased food resources, absence of natural enemies that previously acted to control their numbers, lack of competitors or a combination of these and other factors. Introduced plant species can transform landscapes, as have the kudzu vine (*Pueraria lobata*) in south-eastern USA and the prickly pear cactus (*Opuntia vulgaris*) in Australia. Exotic species of animals and fish can also become highly destructive if released into new habitats, as seen with rabbits in Australia and Nile perch in Lake Victoria in Africa. Invasive exotic species, whether they be plants, animals or fish, can have major effects on the indigenous flora and fauna of natural habitats. Introduced bracken and gorse in New Zealand, for example, have devastated much of the indigenous flora.[12]

As more transgenic crop plants are grown in commercial situations, and experience is gained through the monitoring of genetically modified organisms in the environment, the fears of catastrophic ecological effects are receding to some extent. However, real concerns exist and have to be addressed by careful risk assessment and measures to ensure that any ecological risks are minimized.

Transgenic crop plants could themselves become more invasive and establish themselves as weeds in other crops. This is of particular concern when herbicide resistance is engineered into plants. Parts of transgenic crops could remain in the soil and grow in the following year, within subsequent crops in the same field, where they would be difficult to kill because of their herbicide resistance. Vigorous transgenic crop material might also become displaced into natural habitats, by whatever reason, and threaten wild populations of related plants through competition.

In the UK a research group was established at Silwood Park, Imperial College, London,[13] to study the invasiveness of a transgenic crop, oilseed rape (*Brassica napus*). This crop is known to colonize non-agricultural land. Plants were grown in three climatically distinct regions, in four habitats, over three growing seasons. Unmodified control plants were compared with antibiotic-resistant (kanamycin) and herbicide-resistant (glufosinate ammonium) transgenic plants. The study concluded that there was no indication that genetic engineering for kanamycin or herbicide tolerance increased the invasiveness of oilseed rape.[14]

Genetic modification of crops could also lead to adverse effects on beneficial species. The impact of transgenic crops on pollinating insects is being studied in a three-year collaborative project in France, Belgium and Britain, which was begun in late 1996. Researchers are studying bee pollination in crops of oilseed rape engineered to produce protease inhibitors against insect pests. In experiments with bees that were fed high levels of protease inhibitors in sugar solutions, it was found that they had trouble learning to distinguish between the smells of flowers. In large areas of transgenic rapeseed, therefore, bee behaviour could be adversely affected.[12]

The risk of transgene spread

However, a potentially more serious ecological threat than transgenic organisms themselves becoming weeds or pests is that transgenes will spread, through breeding with wild relatives, to produce offspring containing the introduced gene.

Crop–weed hybridization between transgenic herbicide-resistant oilseed rape (*Brassica napus*) and its wild relative *Brassica campestris* was shown to produce transgenic weed-like plants, with the appearance of *B. campestris* and high fertility, as early as the first-backcross generation.[15] These transgenic weeds, resistant to the herbicide glufosinate ammonium, were found at the experimental site the following spring, among plants germinated from seeds shed at the time of the previous year's harvest.[16] The risk of transgene spread is also increased by pollen from oilseed rape being able to fertilize plants up to 2.5 km away.[17] Gene transfer from radish (*Raphanus sativus*) to wild weedy relatives has also been detected over distances up to 1 km away from the crop.[18] Hybrids from these crosses showed 'hybrid vigour', and produced significantly greater amounts of seed than normal.[19] More recently, transgenes from herbicide-resistant oilseed rape were found to persist for several generations in hybrids of rape and wild radish, the latter being a common weed around agricultural land. These hybrid weeds were herbicide resistant.[20] A high frequency and range of gene flow was also demonstrated from a plot of transgenic potatoes. Transgenes were found in 72 per cent of unmodified potato plants in the immediate vicinity of transgenic plants, and at levels of around 35 per cent of unmodified plants grown up to 1,100 metres away.[21] These studies demonstrate the ease with which transgenes could spread to wild relatives and how quickly genes could become established in the wild if they proved to be neutral or advantageous.

If foreign genes find themselves in wild relatives of crop plants, through crop–weed hybridization, they may be subject to gene regulation different from that for which they were designed. Unpredictable outcomes may result from uncertain gene expression, uncertain interactions with other genes, uncertain effects of genes through sex and recombination, or the general ecological uncertainties to which plant populations are subject. It is therefore desirable to release sterile transgenic plants whenever possible, so that pollen exchange with wild relatives is unproductive. Plants modified for herbicide resistance should also remain susceptible to at least one major group of herbicides. Crop seed is likely to be a more important route of gene spread than pollen. Seed travels long distances from seed merchants, to farmers and processing factories, giving plenty of opportunity for spillage in transit.[5] The spread of transgenes in weed species could have effects on biodiversity if the transgenic weeds were to become particularly invasive. However, it should be noted that other factors, such as habitat destruction by human activity, pose a far bigger and more immediate threat to biodiversity.[14]

To date, only familiar crops with small genetic changes have been released to the environment. However, transgenic plants are currently being developed for drought tolerance, nitrogen fixation and other more complex characters. The escape of transgenes for these characteristics to the wider environment could lead to plants becoming highly invasive. The extension of the range of a species produced by increased drought tolerance might lead to profound ecological changes. No transgenic species can be perceived in isolation in such cases, as whole habitats are potentially affected.

Gene transfer from plants to viruses has already been described, but gene transfer can also occur from plants to other groups of micro-organisms. Genetically engineered oilseed rape, black mustard, thorn-apple and sweet peas, all containing antibiotic-resistant genes, were grown together in an experimental study with the fungus *Aspergillus niger*. In each case, the fungus incorporated the antibiotic-resistant gene.[22] Transgenic crops commonly contain antibiotic marker genes. The horizontal spread of these transgenes to micro-organisms could have a range of ecological effects. They could, for example, theoretically pass from micro-organisms back into other species of plants or animals, including farm animals, where enhanced resistance to veterinary drugs may result.

Transgenic animals are easier to contain than either micro-organisms or plants, but in future may be a cause of ecological concern. Rabbits are used as experimental animals, for example in studies of protein production in their milk, and may themselves be farmed in future as bio-reactors for pharmaceutical drugs. Rabbits have a history of causing ecological problems and would need effective containment in case genetic modifications had unforeseen effects in the wild. Transgenic fish are not domesticated and most species will survive in the wild, where they have a high reproductive potential. Pacific salmon, genetically engineered so that they no longer annually migrate from salt water to fresh water, for example, could pose a significant ecological risk. Instead of returning to their native streams to spawn, these salmon live and feed in the ocean and therefore increase their growth rate and subsequent economic value. If these salmon escaped from fish farms and displaced their wild counterparts, there could be major disruption to ecosystems of rivers in north-western USA. This illustrates how a small genetic modification could potentially have a large ecological impact.

Genetically modified organisms themselves can be contained or 'disabled'. However, it is transgenes that will spread to cause potential ecological problems. Genes may have their own agenda. Richard Dawkins proposed in his influential book *The Selfish Gene* that the gene was the fundamental unit of natural selection.[23] This led to the view that ephemeral organisms are just machines, created by genes, for the production of more genes. The genome, as previously noted, is more fluid and dynamic than originally supposed. The presence of mobile genetic elements and the predisposition of cells to incorporate foreign DNA, for example from viruses, contribute to an uncertainty about how transgenes may behave in a genome. Chaos theory suggests that complex behaviour is intrinsically unpredictable.[24] The gene-centric view therefore leads to a pessimistic picture of how controllable the human manipulation of complex genomes can be. The escape of transgenes to the wider environment is bound to happen eventually. As the chaos theorist says in *Jurassic Park*, a fable about the dangers of unregulated genetic engineering, 'life finds a way'.[25]

The question of who should be held responsible for ecological damage was debated within the European Union in July 1997, when the European Parliament declared that companies should be held liable for damage to the environment or health that results from the release of genetically modified organisms. The escape of introduced transgenes to the wider environment has a certain inevitability. It is a question of what happens when, rather than if, transgenes get into species or varieties that they were not designed to be in. Careful forethought during the development of genetically modified organisms can go some way to ensuring the resulting ecological consequences are minimal, although the nature of the genome means that uncertainty will always exist.

Notes

1. *Science* 261: 1271, August 1993.
2. Royal Commission on Environmental Pollution, 1989.
3. Rissler and Mellon, 1996.
4. Crawley, M., 1996, 'The day of the triffids', *New Scientist*, 6 July 1996, pp. 40–1.
5. Doyle et al., 1995.
6. Lindow and Panopoulos, 1988.
7. Wheale and McNally, 1990.
8. Juma, 1989.
9. Bishop et al., 1988.
10. Tepfer and Jacquemond, 1996.
11. *New Scientist*, 16 August 1997, p. 4.
12. Tudge, 1993.

13. PROSAMO (Planned Release Of Selected And Modified Organisms), funded 1990–93 by Imperial College, the British government's Department of Trade and Industry, and a consortium of biotechnological firms.

14. Crawley et al., 1993.

15. Joergensen and Andersen, 1994.

16. Mikkelsen, T. R., Andersen, B. and Joergensen, R. B., 1996, 'The risk of crop transgene spread', *Nature* 380: 31, 7 March 1996.

17. Timmons, A. M., O'Brien, B. T., Charters, Y. M. and Wilkinson, M. J., 1994, *Scottish Crops Research Institute Annual Report 1994*.

18. Klinger, Arriola and Ellstrand, 1992.

19. Klinger and Ellstrand, 1994.

20. *Nature* 389: 924, 30 October 1997.

21. Skogsmyr, 1994.

22. Hoffmann et al., 1994.

23. Dawkins, 1976.

24. A discussion of chaos and complexity is, unfortunately, outside the scope of this book. However, good introductions are available, for example: Gleick, J., 1988, *Chaos: Making a New Science*, Sphere Books, London; M. Mitchell Waldrop, 1992, *Complexity: the Emerging Science at the Edge of Chaos and Order*, Viking, Harmondsworth. Also of interest in this context is the work of Stuart Kauffmann (Kauffmann, S. A., 1993, *The Origins of Order: Self-Organization and Selection in Evolution*, Oxford University Press, Oxford), who demonstrated how order crystallizes spontaneously from chaotic systems. He has applied his theories to gene regulation, to show that 'random genetic programs, defying our earlier prejudices, can exhibit very great order'. The implications are that selection is complemented by self-organization in explaining the workings of the genome. The mechanisms driving our genes are still incompletely understood.

25. Crichton, M., 1991, *Jurassic Park*, Arrow Books, London.

8. Risks to human health

Genetically modified foods are unlikely to present direct risks to human health. However, the unique nature of these foods, in that genes are transferred between species during their production, and the possible unpredictable outcome of transgene effects, warrant that these foods are carefully monitored. There have been two main areas of concern: a) the possibility of allergic reactions to genetically modified foods, and b) the possibility that bacteria living in the human gut may acquire resistance to antibiotics from marker genes present in transgenic plants.

Allergens

Allergy is an imbalance in the immune system, also known as immediate hypersensitivity. In a normal immune response, a foreign substance called an antigen triggers the production of antibodies. Antibodies are specific to particular antigens: antibody molecules fit around antigens and destroy them. Antigens from, for example, disease-producing microorganisms prompt the build-up of antibodies that act as a defence against further attack. In an allergic reaction, however, inoffensive substances or allergens can trigger a cascade of inappropriate defence mechanisms. Allergy is a blanket term that encompasses various types of immune reactions and pathological states, including asthma, hay fever, eczema and, most seriously, anaphylactic shock. Allergens may be inhaled, as dust or pollen, injected, picked up by contact or eaten. Food substances have been known to cause inhalant allergy. For example, allergens in dust from soybeans, being unloaded from a ship, combined with air pollution to cause an asthma epidemic in Barcelona in 1987. Food allergies are, however, more commonly caused by consuming substances that trigger reactions in the digestive system, such as vomiting and diarrhoea, or reactions affecting the whole body, such as the skin conditions eczema and urticaria.[1]

Between 1 per cent and 2 per cent of the populations of most Western countries have an allergic reaction to certain food types. The most common allergies are against milk, eggs, peanuts and other nuts, shellfish, molluscs, fish, soya and cereals. Fruits and vegetables, such as strawberries, apricots, carrots and celery, can also cause allergic reactions in small groups of sensitive people. The genes coding for many of the proteins involved in allergic reactions have been identified, and these can therefore be avoided when genetically modifying organisms for food. Genetic engineering can also be used to remove allergic proteins from foods. For example, in Japan, a protein that provokes allergic reactions has been experimentally removed from rice.

Transferring genes to a food product may alter the degree to which that product causes allergic reactions in sensitive people. Most new substances present in food due to genetic engineering are likely to be proteins present in trace amounts. Unfortunately, trace amounts of an allergen are sufficient to trigger physiological reactions. Soybeans modified to make them more nutritious, by the company Pioneer Hi-Bred, used a gene from brazil nut that coded for methionine, one of the few nutrients that soybeans lack.[2] However, the process also transferred a major food allergen from nuts to soybeans, causing them to invoke similar allergic reactions to the brazil nuts themselves.[3] This was shown when serum and skin tests were done with volunteers known to be allergic to brazil nuts.

In 1992, the US Food and Drug Administration (FDA) stated that genetically engineered foods must be tested and labelled for allergy sensitivity if they have been created using DNA from any foods known to cause an allergic reaction. This ruling would enable further problems like the transfer of the gene from brazil nut to be identified; but it does not apply to most genetically modified foods. Crops modified with genes from bacteria, for example, are not required to be tested for allergens under the FDA ruling. It is a policy that appears to favour industry over the protection of consumers.[4]

Soya products are used in infant formula for babies that are allergic to milk products, as well as in non-milk dairy products for adults. Herbicide-resistant soybeans, containing genes from bacteria, are beginning to be widely used in such products. The risks of allergic problems from Monsanto's Roundup Ready™ soybeans will be small, because the soybeans are not modified for changes in biochemical composition. However, when genetic engineering is carried out to change the nutritional quality of foods, changes in protein content are often made. These products will need to be carefully monitored for possible allergic problems.

Pollen from transgenic crops, collected by bees, may lead to allergic problems for consumers of honey. A study conducted at Leicester University, for the Ministry of Agriculture, Fisheries and Food (MAFF) in Britain, showed that transgenic pollen proteins could remain active in honey for several weeks. The risks may be small, but as increasing numbers of genetically engineered crops are grown in the countryside, all such potential risks need to be seriously addressed.[5] This case illustrates how a genetic modification in one organism can affect a completely unrelated foodstuff.

Allergies are on the increase in industrialized countries: for example, asthma has increased by 30 per cent, and skin allergies have more than doubled since the 1970s.[1] The reason for the increase lies with modifications to environment and lifestyle. Hundreds of chemicals put into the environment by human activities are known to cause allergic reactions. Identifying which allergen is responsible for a particular symptom is a difficult process. Food additives, however, have been linked to allergy, food intolerance and child hyperactivity. A European Community report in the 1980s estimated that between 0.03 per cent and 0.15 per cent of the population were intolerant to food additives. Tartrazine (E102), a food colouring, was one of the first food additives to be positively linked to allergic problems.[6] As food becomes more synthesized, through the use of genetic engineering, the problems of allergic reactions from foods are unlikely to diminish.

Antibiotic-resistant micro-organisms

Marker genes are routinely integrated into transgenic crops to select transformed plants from untransformed plants (see Chapter 2). A common way of doing this is by transferring genes that confer antibiotic resistance into plants. Micro-organisms produce antibiotics as a defence against invading bacteria. This has selected for bacteria with antibiotic resistance mechanisms. Genes for antibiotic resistance from these bacteria can be isolated and transferred to plants. For example, Calgene's Flavr Savr™ tomatoes contain a gene conferring resistance to kanamycin and neomycin, while Ciba-Geigy/Novartis' *B.t.* maize has a gene that confers resistance to ampicillin. These antibiotic marker genes are situated next to genes for the desired character in transgenic plants and are therefore linked to them. When plant material is treated with antibiotic, the genetically transformed material is therefore selected from untransformed material.

The antibiotics used to select transformed plant material are also in clinical and veterinary use in many countries. A number of studies have claimed that antibiotic resistance genes present no risks to humans or animals.[7] However, there is concern that antibiotic resistance genes will be transferred to bacteria living in the guts of humans or animals.

This could reduce the efficiency of antibiotic drug treatments. The Advisory Committee on Novel Foods and Processes (ACNFP), for example, advised the UK government to vote in the European Union (EU) against the authorization for placing Ciba-Geigy's *B.t.* maize on the market in 1996. They claimed that the presence of an intact gene for antibiotic resistance posed an unacceptable risk, because of possible transfer to gut microflora in animals and humans, particularly through the use of the unprocessed product in animal feed.[8] The EU eventually gave the go-ahead for the marketing of this maize, but several member states had reservations about the potential risks the antibiotic marker gene posed to animals and humans.

It is generally accepted that the digestive system acts as a natural barrier to DNA, as the acidic conditions in the guts of animals and humans breaks down DNA. Most DNA is certainly broken down in this manner. However, some DNA has been shown to survive in both the gut and blood system of animals, for example, in studies where DNA has been fed orally to mice.[9] Therefore, ingested foreign genes in foods could theoretically transfer to gut bacteria. Selection pressure would favour bacteria containing antibiotic resistance genes during any course of antibiotic treatment, causing them to become predominant in the gut. This could make certain clinical or veterinary antibiotics less effective. Different genetically modified foods are likely to carry different risks of spreading antibiotic resistance. Foods with live micro-organisms containing foreign DNA carry the most risk, as frequent conjugations between bacteria lead to much exchange of genetic material. The ACNFP recommended that such foods, containing for example lactic acid bacteria, should not contain antibiotic-resistant marker genes.[10] A lower risk is attached to plant material that is to be cooked, or to uncooked seeds of modified plants, while transfer of antibiotic resistance is least likely from highly processed foods.

Ciba-Geigy claimed, in defence of their *B.t.* maize, that even if ampicillin genes were transferred to micro-organisms in human or animal guts, this would have no serious clinical or veterinary implications, because high levels of resistance already exist among human and animal pathogenic organisms.[11] In the case of humans, this is a result of the intensive use of ampicillin in medicine. The intensive use of antibiotics in general has led to widespread resistance developing in disease-causing micro-organisms. It has been suggested that the genes for antibiotic resistance have been co-administered with antibiotics from the start of their clinical use,[9] and that even genes for multiple antibiotic resistance were widespread in the 1950s, although they have become much more common in recent years. However, evidence now suggests that antibiotic marker genes could result in higher levels of antibiotic resistance, and Ciba-Geigy's claim that this contribution is unimportant is disingenuous, as two wrongs do not make a right.

In the case of livestock, the risk posed by antibiotic marker genes has been compared to the more immediate risk of resistance developing due to antibiotics being directly fed to livestock.[12] Low-level doses of antibiotics have been given in livestock feed for over fifty years, to keep animals in good health. The more widespread use of antibiotics in cattle feed, however, dates from the mid-1980s. Antibiotics are given to improve feeding efficiency, so animals need less food to reach marketable size. Kanamycin and other antibiotics, used as selectable markers in genetically modified plants, are therefore already widely used in livestock feed. The increased use of antibiotics in cattle feeds has coincided with an increase in outbreaks of poisoning from strains of *E. coli* and *Salmonella typhimurium*, while evidence is growing that antibiotic-resistant bacteria can spread from animals to humans. This is of concern as the same antibiotics are often used to treat animals and humans. The USA, for example, permits the use of penicillin and chlorotetracycline as growth promoters in animal feed, even though they are routinely used to treat humans.[13] In one case, bacteria resistant to the antibiotic vanomycin were found in the pus from a wound, caused by a fork-lift truck, of a worker in a chicken-packing depot, and chickens were cited as the likely source of the antibiotic-resistant bacteria.[14]

The assertion that antibiotic marker genes will make only a small contribution to increasing antibiotic resistance in gut micro-organisms is entirely speculative. The contribution could be significant – for example, if *B.t.* maize becomes widely used in livestock feed then kanamycin and related antibiotics may become less effective at treating disease in cattle. It may also be significant in certain clinical cases. In 1997 proposals to extend the sale of Zeneca's transgenic tomatoes, to include whole and canned tomatoes, provoked concern because of the presence of a kanamycin-resistance gene.[15] Kanamycin is one of the last-resort drugs for treating multi-resistant tuberculosis, a disease that is on the increase. Ampicillin is also widely used in combined antibiotic therapies. A report by the ACNFP in July 1994 recommended that safety evaluation of antibiotic markers should include assessments of the clinical use of the antibiotic, the likelihood of transfer of resistance genes into, and expression in, gut micro-organisms, and the toxicity of gene products.[10]

Although seen by many as unnecessary,[7] the development of alternatives to antibiotic resistance marker genes is desirable in many transgenic crops destined for human consumption. The ACNFP, for example, in their report called for further research on the development of alternative selectable marker systems.[10] Some alternatives that currently exist include enzymic excision, to cut out marker genes in micro-organisms, and herbicide resistance markers in crop plants. Some alternative marker systems will present their own problems and a case-by-case analysis to choose the most appropriate marker system is desirable.

Notes

1. Frossard, 1991.
2. Townsend et al., 1992.
3. Nordlee et al., 1996.
4. Nestle, 1996.
5. *Sunday Telegraph*, 18 May 1997, p. 7.
6. Millstone, 1986.
7. Bryant and Leather, 1992; Nap et al., 1992.
8. *AgBiotech: News and Information* 8 (9): 159N. The various groups of antibiotics have different modes of action against bacteria. For example, ampicillin and other penicillins (the beta-lactamases) inhibit the synthesis of a layer in bacterial cell walls; kanamycin and neomycin (aminoglycosides) disrupt protein synthesis by binding to bacterial ribosomes. The ACNFP were particularly concerned with the gene in *B.t.* maize conferring resistance to antibiotics in the beta-lactamase group.
9. Webb and Davies, 1994.
10. Advisory Committee on Novel Foods and Processes (ACNFP), 1994, *Report on the use of antibiotic resistance markers in genetically modified food organisms*, July 1994, ACNFP, MAFF, UK.
11. Ciba Seeds, 1996, *Documentation on Bt-maize from Ciba Seeds.*
12. *Nature* 384: 304, 28 November 1996.
13. Bonner, J., 'Hooked on drugs', *New Scientist*, 18 January 1997, pp. 24–7.
14. *New Scientist*, 12 April 1997, p. 13. See also *New Scientist*, 6 December 1997, p. 5.
15. *Daily Mail*, 3 March 1997, p. 9.

9. Some ethical and moral issues

The application of genetic engineering to food production has raised a number of ethical and moral concerns. In this chapter, three issues are examined: ethical concerns about the transfer of particular genes, whether genetic modification increases the suffering of animals, and whether life can morally be owned.

Ethically sensitive genes

Consumers may have particular ethical objections to genetically modified foods. The Committee on the Ethics of Genetic Modification and Food Use,[1] established by the British government and chaired by John Polkinghorne, first reported in 1993. It identified three areas of potential ethical concern: i) the transfer of human genes to animals used as food; ii) the transfer of genes from animals whose flesh is forbidden to certain religious groups to animals that are permitted as food; and iii) the transfer of animal genes to crop plants, which might then become unacceptable to vegetarians. Broader ethical issues were outside the committee's remit.

These 'ethically sensitive genes' were of relevance only to a few foods at that time. However, the report aimed to clarify the situation for future decision makers. To this end the committee took the following facts into consideration: i) because of gene cloning and the replication process, the majority of genes are copy genes and not the original DNA; ii) genes take on their biological role only in the context of the organism in which they operate; and iii) in some biotechnology, none of the original transgene material, or cloned copies of it, enters the final food product.[1] Many molecular geneticists consider that all transferred genes are effectively synthetic copies of the originally isolated gene, as the process of cloning results in a massive dilution of the original gene. However, the committee's guidelines aimed to identify any 'moral taint' that might be attached to genetically modified foods.

The possible transfer of human genes to food is an issue that has already occupied the UK government's Advisory Committee on Novel

Foods and Processes (ACNFP). The production of transgenic animals is still largely a hit-and-miss affair and, for every successfully modified animal, many unsuccessful ones are produced. Founder animals incorporate human genes and produce pharmaceutical drugs, but unsuccessfully transformed animals might also contain human genes that are not expressed properly. These animals are of no value for drug production, although they are of value as normal farm animals. The ACNFP felt that all animals originating from a genetic modification programme should be treated as potentially modified, in case some undetected human gene had been transferred.[2]

The use of genes from animals forbidden by certain religious groups is a complex issue as different religious groups, as surveyed by the Ethics Committee, had different outlooks on genetic engineering, although all agreed that the purpose of gene transfer was the key to their ethical stance. Muslim groups felt that transferred genes retained their origin and that, therefore, a gene taken from a cow will always be a cow gene, while Jewish groups thought genes took on the nature of the organism into which they had been transferred. Muslim groups saw a clear distinction between improving species through traditional cross-breeding methods and genetic engineering, while Christian and Jewish groups were more likely to see humans as people who had been granted power to manipulate nature, with genetic engineering being just another part of this power.[2]

The production of transgenic plants raises fewer ethical concerns than does that of transgenic animals. However, genes from bacteria, fish and animals have been integrated into crop plants. The latter case raises ethical concerns for vegetarians. Many vegetarians would find the consumption of plant foods containing animal genes unacceptable. However, the Vegetarian Society has approved cheese produced using yeast containing a chymosin gene. This 'vegetarian cheese' is an alternative to cheese made with chymosin from calf rennet. A gene from the flounder that protects it against freezing has also been experimentally integrated into tomatoes to prevent frost damage. The majority of genes transferred to transgenic plants, however, are derived from bacteria or other plants.

Animal welfare

The application of genetic engineering to modify farm animals has raised concerns about animal welfare. Although transgenic animals have been produced since the mid-1980s, issues such as how to minimize suffering and how to regulate their production are still unresolved.

Colin Tudge, in his book *The Engineer in the Garden*, argued that genetic engineering raised no radically new welfare or ethical issues.[3] This is because traditional livestock breeding has already provided improvements in food production that are detrimental to animal health. These adverse effects have been exacerbated by the widespread use of growth hormones and the use of intensive rearing units. Genetic engineering does, however, give the breeder more scope for making genetic change. Characteristics favourable for food production could be taken to further extremes – for example, cows could have even bigger udders. The udder size of cows has been increasing through traditional breeding methods, with a corresponding increase in mastitis, a painful udder infection (see Chapter 3). Higher productivity could be obtained from transgenic animals, but it is likely that these animals would be more prone to stress and disease.

Transgenic animals have already been shown to develop complications due to the effects of introduced genes. A gene for bovine growth hormone has been transferred to sheep and pig embryos, but the long-term elevation of levels of this hormone proved detrimental to the animal's health.[4] The most extreme reported problem to date was the case of the 'Beltsville pigs'. A gene for human growth hormone was integrated into the genomes of pig embryos at the USDA research station in Beltsville, USA, in the mid-1980s, with the aim of increasing growth rate. The 'Beltsville pigs' developed severe arthritis, had spinal deformities, became blind or cross-eyed and were impotent. This study has been widely quoted by opponents of genetic engineering and resulted in a great deal of negative publicity for those wanting to develop transgenic animals for human food.[5] Human genes have also been experimentally incorporated into cows, sheep, mice, rabbits and fish.

Concerns about animal welfare should probably be addressed to all animals reared under intensive farming conditions, rather than singling out those with particular genetic modifications.[3] The current laws concerning the welfare of transgenic animals are often vague. In the USA, the welfare of all experimental farm animals is covered by the US Animal Welfare Act. Once an animal has been genetically modified to produce a protein for medicinal use, however, its welfare is largely regulated by the Food and Drug Administration, under the same laws that govern the production of drugs by bacteria.[6] The US regulations allow the production of herds of transgenic animals, even though no studies have been done on the long-term effects on animal health. In Britain, work on transgenic animals is covered by the Animal Scientific Procedures Act of 1986. Licences are awarded where the benefits to humans outweigh the costs, that is the suffering of animals. The costs to animals have been increasing in recent years. The Royal Society for the Protection of Animals has called for a halt to genetic engineering when it is detrimental in any way to the health or well-being of animals. However, it is often difficult to determine the suffering caused to animals by genetic manipulation.

Much of the public's objection to genetic engineering has a moral basis, according to opinion polls.[7] Milk and meat are already produced in sufficient quantities in industrialized countries, and therefore transgenic animals are being raised solely for profit. If additional suffering caused to animals by genetic modification is compared to the benefits gained, then the use of transgenic animals for food production can be viewed as ethically questionable, because the ends do not seem to justify the means.

The development of transgenic animals for the production of therapeutic drugs is perceived as having clearer benefits, fulfilling definite medical needs.[7] However, these therapeutic drugs could also be produced in bacteria using biotechnological processes. In addition, the public find certain medical uses of transgenic animals morally unacceptable, including certain research and xenotransplantation work.[7] Xenotransplantation is the transfer of animal organs to humans. World demand for organs is growing by 15 per cent a year, while organ donation rates are static. Therefore, this work has enormous potential benefits. Imutran[8] are developing 'humanized pigs', engineered with human genes, so that transplanted organs are not rejected.[9] Pig tissue has been successfully transplanted into people. However, recent concerns about a pig retrovirus, which has infected human cells in the laboratory, may severely limit the application of xenotransplantation.[10]

Is DNA life?

Many people object morally to the patenting of life forms. However, to the companies involved, the awarding of patents is essential for the protection of their investment in research and development. Patent law is seen as necessary for the commercial advancement of the new biotechnology.

The moral objection to patenting is greatest for animals, and particularly so for human genetic material. The first animal to be patented was the OncoMouse™ in 1988.[11] This transgenic mouse develops cancer a few weeks after birth. The patent was awarded to Harvard University for its use in testing for carcinogenic effects of drugs and other chemicals. The animals are more sensitive to carcinogens, which makes screening for potential cancer-causing chemicals a much quicker process. Du Pont funded the initial research and has reaped considerable economic rewards. The European Patent Office (EPO) initially refused the OncoMouse™ a patent, as it did not feel that the benefits to humans outweighed the suffering caused to the animals, a clause found in the 1986 Animals Act, in the UK, and other European legislation. However, this objection was overturned on appeal. This set a precedent and a large number of transgenic animal patents have subsequently been awarded. For example, by 1995 over two hundred and fifty patents had been awarded on strains of transgenic mice modified for different gene defects.

The pressure group Compassion in World Farming has opposed the patenting of transgenic animals on animal welfare grounds. For instance, a patent issued to the Australian company Bresagen, covering transgenic pigs that produced extra growth hormone, was opposed because the animals suffered through arthritis, gastric ulcers and diabetes.[12]

The EPO granted its first patent to a human gene (H2-relaxin) in 1991.[13] In 1995, a group of MEPs tried to overturn the patent, arguing that patenting a human gene is tantamount to patenting human life and is therefore immoral. The EPO defended its decision and concluded that 'DNA is not life'. The EPO in its deliberations saw no moral distinction between patenting human genes and human proteins.[13] Concerns about the patenting of human genetic material now centre on the Human Genome Project, which aims to sequence the approximately hundred thousand genes on the human genome by around the year 2005,[14] and the controversial patenting of genes from indigenous peoples, for example tribes in Papua New Guinea, by the National Institutes of Health in the USA.

Patent lawyers are having to deal with an ever increasing number of patent applications concerning genetic modifications, which are seen by the general public as having moral implications. It was never intended that patent lawyers become arbiters of what is right and wrong, and for lawyers the moral debate is a distraction from their main objectives. Patent law does, nevertheless, consider ideas of morality. The European Patent Convention of 1973 excludes from patentability any invention 'the publication or exploitation of which is contrary to morality or *l'ordre publique*' (Article 52(a)). This was incorporated into the Patents Act of 1977, to prevent the patenting of inventions that encourage offensive, immoral or antisocial behaviour. It was used unsuccessfully by both the opponents of the OncoMouse™ patent, on the grounds that it encouraged cruelty to animals, and by opponents of the patenting of herbicide-resistant crops, on the grounds that they encouraged the indiscriminate spraying of crops with agrochemicals.

On 16 July 1997, the European Parliament voted to accept a draft of the Directive on the Legal Protection of Biotechnological Inventions, which allows the patenting of living organisms and genes, including human cells and genes. The directive endorses the patenting of life forms for the first time in European law. In principle, when the directive becomes law in 1998, patents should only be awarded to inventions, where an inventive step in production can be shown, and provided procedures are within defined ethical limits. An amendment led to the establishment of a bioethics committee to scrutinize patent decisions. Another amendment prevented the patenting of whole human beings.[12] Critics claimed, however, that the distinction between inventions and discoveries had become blurred by the wording of the directive, leaving the door open for private companies to patent genes and gene sequences as they occur in nature.

The directive also concerned itself with animal welfare issues, by including a clause that excludes patents on processes for modifying animals 'which are likely to cause them suffering or physical handicaps without any substantial medical benefit to man or animal'. The inclusion of the word 'medical' meant that agricultural benefits will not necessarily outweigh animal suffering. This could be grounds for rejecting patents on, for example, transgenic animals modified for faster growth.[12]

As with any other technology, genetic engineering can be used for the greater or lesser good of society. It is for society to use ethical and moral judgements to produce legislation that regulates this technology, so that undue suffering is not caused to animals, and so that morally acceptable applications, bringing genuine benefits to people, are developed.

Notes

1. *Report of the Committee on the Ethics of Genetic Modification and Food Use*, HMSO Books, London, 1993.
2. Aldridge, 1994.
3. Tudge, 1993.
4. Pursel et al., 1987.
5. Penman, 1996; *Guardian, G2*, 5 March 1997, p. 4.
6. Gordon, M., 1997, 'Suffering of the lambs', *New Scientist*, 26 April 1997, pp. 16–17.
7. *Nature* 387: 845–7, 26 June 1997.
8. Imutran is based in Cambridge, England. It was taken over by Sandoz in 1996.
9. *New Scientist*, 31 August 1996, p. 10.
10. *New Scientist*, 1 March 1997, p. 6.

11. Aldridge, 1996.
12. *Nature* 388: 311, 24 July 1997.
13. *New Scientist*, 28 January 1995, p. 8.
14. Five per cent of funding within the Human Genome Project is now diverted to ethical issues, coordinated by the project's ethical, legal and social implications working group.

10. The lucrative art of patenting

Multinational companies hold patents on a wide range of genetically modified organisms, and on the techniques that produced them. Patents give them exclusive intellectual property rights on organisms, genes or processes for up to twenty years. In the area of food production, the main profits are generated from the sale of patented transgenic crop seed. This is more lucrative than it first seems, as explained in this chapter, because of gene-licensing agreements, international trade agreements and the extension of intellectual property rights so that they apply globally. Farmers, particularly in the Third World, may be adversely affected by the patenting of crop seed.

Patenting plants

Patent rights are granted in exchange for disclosure of information. This information prevents other people from infringing the patent, allows the public access to the patent's secret on its expiry, and prevents further patents being issued on existing inventions.[1] Multinational companies view patents on the products of biotechnology as essential, to protect their substantial investments in research and development. Companies holding patents can license patent rights out in exchange for royalty payments or licence fees. Royalties are payable for use of transgenic crop seed, for example, and on all seed subsequently produced from these transgenic plants, for the duration of the patent.

Historically, the patent-like protection of life forms has been treated separately to patents involving inanimate objects. Plant Breeders' Rights (PBRs) were awarded to plant breeders to safeguard new crop varieties. In the USA, these rights were first granted in 1930 with the passing of the Plant Protection Act (PPA), to cover asexually reproducing varieties. These new varieties had to be distinctive to fulfil a requirement for novelty. In 1961, the International Convention for the Protection of New Plant Varieties was signed by 18 countries and gave breeders intellectual property rights over new varieties. In 1970, the Plant Variety

Protection Act (PVPA) extended Breeders' Rights to include sexually reproducing varieties. These PBRs protected against the resale of seeds, but permitted the use of seeds by any plant breeder as a parent for future generations and allowed farmers to store and resow seed from the protected crop. Legal obstacles in the past to awarding new varieties full patent status have included the requirement for reproducibility, the requirement for a complete description of the invention, and the fact that the starting materials are a 'product of nature' and not solely inventions arising from the creative power of the human mind.[1]

The requirement for reproducibility has proved a major obstacle in the past for plant breeders because new plant varieties that result from mutations prove difficult, if not impossible, to reproduce, despite the well-established procedures and skills of breeders. Inbred lines will yield patentable plants, however, as first recognized under the PVPA, due to the processes of self-pollination and the selection of desired characters leading to seeds having similar genomes. Inbred lines will therefore be relatively distinct, uniform and stable. Hybrid seed does not breed true and could not be patented. However, this in itself provides an alternative form of protection, as farmers need to buy new seed for each crop.[1] The PVPA allowed control over the inbred lines that are used in the production of hybrids, which guaranteed that hybrids are effectively protected, and parent inbred transgenic lines are similarly patentable. A great deal of investment has gone into the production of hybrid seeds, and most vegetable crops have become hybrid as a result of plant-breeding efforts.[2] Therefore it is the pure-breeding lines of genetically modified crops that are patented, and these are crossed with other varieties to produce hybrid crops containing the transgenes.

Inanimate objects are easier to describe than plants because they do not change over time in a manner that compromises their description. The application of the concept of intellectual property rights to new varieties of plants, however, helped to erode the 'complete description' obstacle to patenting.

In the USA an important test case, Diamond v. Chakrabarty in 1980, paved the way for the granting of patents to life forms, by rejecting the 'product of nature' principle for the first time. In this case, the US Supreme Court decided that a new strain of the bacterium *Pseudomonas* could be patented, after its developers appealed against a decision by the US Patent and Trademark Office (PTO) to reject the patent application. The bacterium was developed to digest oil slicks and the patent was awarded to General Electric. The court ruled that the bacterium was not a naturally occurring species, and therefore was the result of human invention.

The first broad biotechnology patents were awarded between 1980 and 1984 to Stanford University and the University of California for the basic DNA recombinant technology developed by Boyer and Cohen. They used the technology to produce insulin, hepatitis B vaccine and other products in bacteria. The patents covered most of the genetic manipulation techniques in use at that time. To reduce the possibility of challenges to its patent, Stanford University adopted a non-exclusive policy and relatively modest licence pricing. This led to the technology being widely distributed and enabled many other commercially important developments in biotechnology to be made using the techniques. However, critics have argued that taxpayers' money had paid for the Stanford research, the patenting of which has earned the university millions of US dollars, while the costs have been passed back to the taxpayers when they buy the products of biotechnology.[3] The broad nature of the patents, which start to expire in 1997, was resented by many in the industry, but it set a trend for future patent applications.

Major commercial pressures have long existed for crop plants to be awarded patent status. PBRs extended only to the initial sale of seeds, while patent protection means that farmers and plant breeders have to pay royalties to the patent holder for every subsequent generation, produced from the original seeds, throughout the lifetime of the patent. In addition, the PBR system protects only products, while the patent system protects both products and processes, enabling patent holders to earn higher royalties.

The European Patent Convention of 1962 stated that 'essentially biological processes' could not be patented. However, the European Patent Office (EPO), based in Munich, issued patents on transgenic plants and animals on a case-by-case basis for a number of years, starting in the late 1980s. The first patent on a genetically engineered plant in Europe was granted in 1989 to Lubrizol, who sought rights on a technique that modified sunflower, alfalfa and soybean, so that they stored more protein.[4] A draft directive initiated in 1991 by the European Commission (EC) proposed that the products of biotechnology were produced not by biological processes but by human invention and were, therefore, patentable. This directive was voted against by the European Parliament in 1995 because of ethical concerns about patenting life and human genes, as well as concern that it could impinge on plant breeders' and farmers' rights. The EPO, meanwhile, had stopped awarding patents to transgenic organisms.

The EC carefully reworded the directive in the light of previous objections, and the Directive on the Legal Protection of Biotechnological Inventions was eventually approved in draft form in July 1997.

A number of amendments ensured that discoveries, such as a gene in its natural state, could not be patented. Multinationals had lobbied hard and managed to convince a majority of MEPs that a patent directive was necessary to ensure that the European biotechnology industry kept up with that of the USA and Japan, while jobs and the science research base were preserved. This 'fear of Europe being left behind' is a common argument used by those promoting genetic engineering, and will also be met in the debate on marketing approvals for genetically modified food. However, it was US-based multinationals and biotechnology companies, eager to market their products and expand their business in Europe, who lobbied the hardest for the patent directive – the same companies who have the most patent applications pending with the EPO. The directive did not stipulate that patent holders have responsibilities to establish research or production in the region where the patent is granted. However, the directive will also help smaller biotechnology companies in Europe and it cleared up a range of legal uncertainties. It strengthened the industry position by ensuring that patents are more likely to be awarded to genetically modified products for agriculture, but made it more difficult for organizations to object to the patenting of life forms on ethical grounds (see Chapter 9). The EPO had a backlog of around one thousand two hundred patent applications for plants, and around six hundred for animals, by the time the directive was approved.[5]

Disputes between companies have frequently arisen over patenting. The broad nature of many patents, and inconsistencies between patents, set off a round of legal challenges. The outcome of these cases can have major financial consequences for the companies concerned. In 1991, ICI sought to secure a patent on a genetically engineered tomato that remains firm when ripe. Calgene, a rival company, had a patent on a similar plant, the Flavr Savr™ tomato. Both tomatoes were produced using gene-silencing techniques, although ICI had used a sense gene construct and Calgene had used an antisense gene construct (see Chapter 6). ICI wanted to patent the particular stretch of DNA integrated into its tomato, while Calgene had a patent covering its technique for producing transgenic tomatoes.[4] Such cases highlighted the confusion caused when patents are awarded to both genes and processes. This case was followed by another, in 1993, when the US PTO awarded Enzo Biochem rights to antisense technology, used in the production of various transgenic plants. This patent gave the company broad rights to use novel RNAs, called antisense RNAs, to block the activity of specific genes in any crop. Enzo Biochem immediately tried to sue Calgene for infringing this patent by producing crops using antisense technology, including Flavr Savr™ tomatoes, although the challenge was unsuccessful.

In 1995, the US PTO awarded a patent to Mycogen for rights to any method of modifying *B.t.* insecticidal protein genes to make them resemble plant genes. Mycogen is a relatively small company and, rather than enter into a costly legal battle with Monsanto, a much larger company who had developed similar technology, it negotiated for a licensing agreement.[6] Patent conflicts are gradually giving way to co-operative licensing agreements. Many small biotechnology companies are being taken over by multinationals, leaving patents on major crop plants largely in the hands of a small number of companies.

The patent system in the USA, Europe and elsewhere was founded to deal with mechanical inventions, and now has to cope with modifications to biological systems. Patents were usually awarded to individuals or small companies to give them intellectual property rights over their inventions. However, it is now multinational companies that hold patents. In 1990, 50 per cent of the plant patent applications to the EPO were from just eight multinational companies. One-third of applications came from just three companies: Monsanto, Ciba-Geigy and Lubrizol.[7] Multinationals now have a great deal of influence over the patenting process, while patent lawyers are having to make decisions on broad patent applications covering major food crops that have implications for food security.

Species-wide patents

In October 1992, the PTO controversially awarded a patent to Agracetus, then a wholly owned subsidiary of W.R. Grace (although now owned by Monsanto), for rights to all forms of genetically engineered cotton, no matter what techniques were used, or what foreign genes were used, to create the transgenic plants.[8] In 1994, the US Department of Agriculture (USDA) and the law firm of Breneman and Georges mounted challenges to Agracetus' cotton patent. The PTO agreed to re-examine the patent and, by the end of the year, had thrown it out on the grounds that the patent neglected to refer to cotton transformations done by another company, and that the production of genetically engineered cotton had already become 'obvious' to scientists in the field. Meanwhile, Agracetus proceeded with similar patent applications on transgenic cotton around the world. However, India revoked an Agracetus 'species-wide' cotton patent in February 1994, under a clause in the Indian Patent Act of 1970, 'because of its far-reaching implications for India's cotton economy'.[9]

The cotton application, however, was only the first made by Agracetus to secure species-wide patents on important agricultural and industrial crops. Agracetus holds the patent on the Accell gene gun (see Chapter 2), used for introducing foreign genes into crop plants. This proprietary technique helped Agracetus gain its species-wide patents. Agracetus claimed that it needed broad patent protection to protect its investment in developing transgenic crops.[9] In Europe, on 2 March 1994, Agracetus was granted a patent by the EPO for genetically engineered soybeans.[10] This patent covered all genetically modified soybean, irrespective of the techniques employed or the germplasm involved. It in effect amounted to a species monopoly on all genetically engineered soya within the European community for 17 years, the duration of the patent. Similar species-wide soya patent applications were made by Agracetus in the USA and all the other soybean-producing nations. Legal challenges were mounted to this patent, but without success. For example, the non-governmental organization Rural Advancement Foundation International (RAFI) challenged Agracetus' European soybean patent on the grounds that it contravened the *l'ordre publique* section of the European Patent Convention, which excludes from patentability any invention 'the publication or exploitation of which is contrary to morality or *l'ordre publique*'. RAFI claimed that the patent was contrary to public morality, because giving a single corporation monopoly control over genetic research on one of the world's most important food crops is a threat to world food security.[6]

By the mid-1990s, when the company was acquired by Monsanto, Agracetus had succeeded in getting patents covering all genetic manipulation of cotton and soybeans, and had patents pending on rice, groundnuts and maize. Agracetus has submitted patent applications for creating transformed rice plants to patent offices around the world; the first application to the EPO was on 11 May 1992. This patent was believed to seek patent rights to any rice transformed using methods involving immature rice embryos and meristem discs, and to cover both *indica* and *japonica* rice varieties.[9] Agracetus has also used its proprietary Accell gene gun to introduce herbicide resistance into two rice varieties, the US variety Gulfmont and a variety called IR-54 from the International Rice Research Institute (IRRI). The IRRI, based in the Philippines, holds the world's largest collection of rice germplasm and has done much to conserve the biodiversity of this crop. Critics in the developing world are worried by such patent applications. If the insertion of one foreign gene into an IRRI variety renders it the property of Agracetus (Monsanto), then multinationals stand to make massive profits from the endeavours of plant breeders in the Third

World. Crop varieties in general are the culmination of thousands of years of artificial selection. Many farmers and breeders worldwide, over many years, have laboured to get a crop variety to the stage where it can be genetically engineered. Yet one genetic change could render it the exclusive property of a commercial organization.

RAFI has been particularly active in pointing out the implications of 'species' patents on major crops. A few broad patents could be used to control entire areas of research. The soya patent, for example, effectively discourages all other research and development on genetically modified soya. If researchers at a European university, for instance, developed a transgenic soybean plant with a desirable character, they would be in violation of Agracetus' patent if they did not obtain a licence or pay royalties to Agracetus, despite using techniques and germplasm different from those used by Agracetus. Patent laws prevent farmers from saving any seeds from genetically engineered soya, or any other crop covered by a species-wide patent. Soya and cotton, the two crops for which Agracetus has most actively pursued species-wide patents, are non-hybrid and open-pollinated. Therefore, farmers in many parts of the world routinely save a portion of the harvest of these crops for replanting. Although patents apply only in the countries where they are recognized, international trade agreements are now giving patent claims a global force, as discussed later in this chapter. In addition, the countries where patent-holding multinationals are based could be within their rights to prohibit imports of raw materials or finished goods derived from genetically engineered crops covered by species-wide patents but not sanctioned by the patent holder – for example, cotton clothing from India or soypaste from Brazil.[9] Geoffrey Hawtrin, the director-general of the International Plant Resources Institute, in Rome, told RAFI:

> The granting of patents covering all genetic engineered varieties of a species, irrespective of the genes concerned or how they were transferred, puts in the hands of a single inventor the possibility to control what we grow on our farms and in our gardens. At a stroke of a pen the research of countless farmers and scientists has potentially been negated in a single legal act of economic high-jack.[9]

Interfirm cooperation agreements

Multinationals holding broad patents on techniques and major crops are increasingly becoming the norm in agricultural biotechnology. This is resulting in a greater amount of cross-licensing and intercompany arrangements for the use and development of biotechnology.[9] An important trend in biotechnology is the 'interfirms cooperation agreement', whereby companies that have complementary expertise or parallel market interests cooperate on a selective basis. Networks of alliances and joint ventures have, therefore, developed between the major multinational companies, enabling these companies to control key areas of agricultural biotechnology.[7]

A typical cross-linking agreement was made between Calgene and Monsanto in 1996, in which Calgene received a royalty-free licence to use Monsanto technology, i.e. Roundup Ready™ seeds, in combination with Calgene's own proprietary canola genes. In return, Monsanto received a royalty-free licence to Calgene's technology for developing crops. Calgene got a royalty payment from Monsanto based on sales of crop seeds produced using Calgene technology.[11]

After Agracetus was awarded its species-wide European soybean patent, it was subsequently licensed to Monsanto (which now owns Agracetus). Monsanto's Roundup Ready™ soybeans were therefore covered by this cross-licensing agreement. The Accell gene gun technology, patented by Agracetus, is licensed out to organizations wanting to develop transgenic plants. The cost of putting a gene into maize in 1994, for example, was US$20,000 per plant, plus royalties. This would have seemed reasonable to Monsanto at the time, but put the technology beyond the reach of public sector and developing world scientists.[9]

Restructuring and cross-licensing also have major implications for small companies, which may be forced out of business. Arrangements between companies work by a process of give-and-take, and smaller companies may not have enough to bring to the negotiating table. Some commentators even suspect that larger companies may be seeking increased regulation of genetically engineered plants in order to create competitive disadvantages for smaller, more innovative firms. This may explain the initially paradoxical stand of Monsanto, and other major players in the US Biotechnology Industry Organization (BIO), in supporting the Environmental Protection Agency's (EPA) proposal to regulate as pesticides crops engineered for resistance to pests and diseases.[12] The big agrochemical and biotechnology companies know that smaller companies struggle against the artificially high market-entry barriers that are created when excessive regulations are imposed. Until now in the USA, few governmental evaluations have been required of transgenic varieties, compared to the development of a new pesticide, allowing smaller companies to prosper. It is in the interests of the big companies to slow the development of competing products, particularly products from innovating companies that may eat into their markets. Entrepreneurial companies are more likely to fail, because of red tape, long delays and expensive field validations, or they will sell out to companies like Monsanto, as many have already done. The consequence of this may be fewer product choices and higher prices for farmers, food processors and consumers.[13]

The division between academia and industry, or pure and applied research, has also been eroded by biotechnology. New developments in genetic techniques have immediate commercial importance, such as techniques for transferring genes to cereals or techniques for cloning transgenic animals. Both groups are now seeking patents in the same areas. This has led to increasing cooperation between universities and multinationals, and most university researchers in this field now have strong industry links. The independence and non-profitable status of university research is, therefore, becoming increasingly questionable. The patenting process reduces the flow of information from universities, due to commercial considerations, and this is particularly detrimental to researchers in developing countries.

Gene-licensing agreements

The major agrochemical and biotechnology companies have drawn up licensing agreements, which farmers must agree to sign and abide by if they are to use the company's transgenic seeds. In this way, multinationals can control the way their proprietary genes are used. Monsanto's '1996 Roundup Ready Gene Agreement', which farmers had to sign in order to buy Monsanto's Roundup Ready™ soybean seed, gave the company extraordinary influence over the way farmers used the proprietary seed, what inputs could be used to grow the seed, and the access Monsanto were allowed to farms growing the seed.[14] Growers had to pay Monsanto US$5.00 per 50-lb bag 'technology fee', on top of the cost of the premium-priced seed, and to give Monsanto the right to inspect and test their soybean fields for up to three years. Growers also had to use only Monsanto's Roundup Ready™ brand of glyphosate herbicide. The use of any other brand of glyphosate would have been regarded as a violation of the agreement. Growers also relinquished the right to save seed or re-plant patented seed, or sell seed derived from it. If farmers violated the agreement, they must agree 'to pay Monsanto as liquidation damages a sum equal to one hundred times the then applicable fee for the Roundup Ready™ gene, times the number of units of transferred seed, plus reasonable attorney's fees and expenses'. Farmers could, therefore, risk losing their farms for acting according to what until now have been regarded as 'Farmers' Rights'.[14]

The concept of Farmers' Rights, endorsed by the United Nations Food and Agriculture Organization (FAO) in 1989, recognized that farmers have contributed greatly to the conservation of genetic resources and that this should be rewarded in the same way as Plant Breeders' Rights. Farmers' Rights cover the right of farmers to land and tenure, and the right to save and exchange seed. Monsanto's view is that the right of farmers to save or exchange seed from their harvest is an infringement of their patent rights.[14] The arrival of Roundup Ready™ seed was greeted enthusiastically by most, but certainly not all, soybean growers, keen to increase their profits and therefore willing to accept Monsanto's terms. Five to ten million hectares were sown in the USA in 1997. This transgenic soya was also grown outside the USA for the first time in 1997 – for example, in Argentina.

The Roundup Ready licensing agreement will also be applied to other crops engineered with Roundup Ready™ genes, including canola, maize and sugar beet, although the details of the agreement will vary from crop to crop. Monsanto has said that its 1997 licensing agreement with farmers growing Roundup Ready™ soybeans contained fewer conditions than its 1996 agreement in response to farmers' concerns.[14]

The decline of independent seed companies

During the 1980s and 1990s, multinational companies restructured and amalgamated to exploit the potential of biotechnology and genetic engineering, often bringing agrochemicals, seeds and pharmaceutical drugs under the same roof. The seed industry was once a diverse sector, with many family-based firms. However, many seed companies needed to merge to survive during the 1970s. For example, a number of British seed firms, including Sutton's and Cuthbert's, are now part of the giant French horticulture group Vilmorin-Andrieux. Pioneer Hi-Bred are a large specialist seed company, with one of the world's highest single market shares of seed sales, but they are the exception. Most seed sales are through the major multinational agrochemical companies, who have bought up seed companies at a rapid rate and are soon likely to control most of the market.[7]

The control of seeds is the key to the profitable exploitation of genetic improvement in crop plants. Monsanto, for example, has been acquiring seed companies, including Hartz and Dekalb's Hybrid Wheat Program, for a number of years. In January 1997, Monsanto acquired Holden's Foundation Seeds. Holden's produces the inbreds, or parent seed, used by retail companies to produce the hybrid seed sold to farmers. More than 35 per cent of the maize planted in the USA contains genetic material developed by Holden's.[15] This gives Monsanto an excellent platform from which to promote and market its proprietary genes, such as Roundup Ready™ and YieldGard™ *B.t.* genes, in maize.

Multinational companies can claim royalties on genetically modified seeds, and they may start to market them in preference to older and more established seed varieties. Farmers may increasingly find traditional seed harder to obtain from major seed suppliers. In addition, the Seeds Trade Act in Europe makes it illegal to grow and sell non-certified seeds produced from indigenous varieties. With seed companies increasingly controlled by the major agricultural multinationals, critics fear that these companies will come to have a disproportionate influence on seed certification systems in Europe and elsewhere in the world. This may result in a possible bias toward commercial transgenic varieties that carry patent protection. Grass-roots resistance from small organic growers and gardeners has arisen to save seeds of native crop plant varieties that are not listed varieties. Many of these were once commercial varieties that seed companies decided they no longer wanted to stock. Countless varieties are dropped from seed catalogues due to small family-owned companies being taken over. Some of these varieties have characteristics that are being sought in transgenic crops – for example, slow-ripening in tomatoes.

The Henry Doubleday Research Association (HDRA) runs a 'heritage seed library' to preserve the indigenous genetic stock of British vegetables. It estimates that thousands of British vegetable varieties have been lost since the 1970s. This is a trend repeated in many industrialized countries. The HDRA cannot legally sell the seeds in their heritage seed library, because they are not on the National List of Vegetable Varieties, but members join a scheme to have access to the seed varieties. The HDRA is one of the pressure groups worldwide who campaign to preserve the biodiversity of vegetables, against what they see as an erosion of agricultural biodiversity by the industrialization of agriculture.

The GATT and the MAIs: free trade and global rights for multinationals

The General Agreement on Tariffs and Trade (GATT) first came into being in 1948, as a temporary arrangement, largely on the initiative of the USA, who saw 'free trade' as one of the pillars of post-war order. The seven 'Rounds' of the GATT, until 1986, had as their main objective 'the substantial reduction of tariffs and other barriers to trade'. In September 1986, the eighth or Uruguay Round began,[16] and for the first time negotiations were extended to encompass many issues beyond the traditional ones of duties and tariffs, including technological developments.[17]

In the Uruguay Round the patenting of life forms, using the concept of 'intellectual property rights', was extended to the global level. The owner of a patent therefore became the recognized owner of the new life form around the world, and had exclusive patent protection for 20 years. This gave multinationals, holding patents on transgenic seeds, exclusive rights to their use worldwide. National patent laws around the world therefore became subservient to the patent laws of countries where multinationals were awarded their patents. These changes to the GATT added to the already numerous protectionist barriers erected by industrialized countries against Third World countries. The developing world was poorly organized for lobbying within the GATT and subsequently had little participation in its final drafting. The USA, meanwhile, strengthened its capacity to enforce intellectual property rights through international trade. Countries that resisted the GATT legislation risked being subjected to trade retaliation, not only with regard to the commodity or service in dispute, but right across the board.[18] Critics of the GATT pointed out that it favoured multinational companies. Viewed from a Third World perspective, the GATT, with its associated threats of trade retaliation, could be regarded as the method by which industrialized countries retained control of the global economy in the post-colonial age.[19]

In January 1994, the North American Free Trade Agreement (NAFTA) came into effect, eliminating trade barriers between Canada, the USA and Mexico. This was to provide a model for future multilateral agreements. The Uruguay Round itself led to the formation of the World Trade Organization (WTO), the successor to the GATT. The WTO was established on 1 January 1995 to be the legal and institutional foundation of a multilateral trading system. It was set up to provide a forum in which trade relations among countries evolve through collective debate, negotiation and adjudication.[20] Its international trade rulings were signed by 103 countries and any country ignoring WTO rulings is liable to punitive sanctions. The application of free trade agreements is generally stricter than under the previous GATT rulings.[21]

The USA has already defended its interests under the WTO. The US government, acting on behalf of Monsanto, has requested that the WTO declare the European ban on rBST illegal. The US government intends to use the WTO to declare illegal any bans on genetically modified foods exported from the USA that have equivalent composition to non-modified foods. In June 1997, the US government used the WTO to insist that India amend its patent laws. The USA also threatened not to renew its science and technology agreement with India, which would have ended funding for 130 projects.[22] India's existing patent laws, passed in 1970, did not allow patents in the food or pharmaceutical sectors. India had previously revoked an Agracetus species-wide patent, as mentioned earlier, using its national patent laws. After joining the WTO in 1995, the government had drafted an amendment to conform with the WTO's guidelines on intellectual property, but the Indian parliament's upper house had blocked it. The upper house argued that patents would make essential drugs too expensive for the poor and allow seed companies to make agriculture too costly for marginal farmers. However, the Dispute Settlement Board of the WTO ruled against India, meaning that any member of the WTO could take retaliatory action against India with potentially disastrous results for Indian exports. Under this intense pressure, an amendment was made to India's national patent laws.

The globalization of world trade has favoured the multinationals, and their increasing power was consolidated through a series of negotiations held in Paris in 1997: the Multilateral Agreement on Investments (MAI).[23] Initially the EU had proposed that a global investment treaty be developed within the WTO, but the USA feared that the presence of Third World countries would 'water down' any consensus on investments. The USA therefore led negotiations within another forum, the Organization for Economic Cooperation and Development (OECD), where agreement could be reached among the 29 industrial countries that are its members.[24] The MAI aims to establish a set of global rules for investment that will grant multinationals unrestricted rights and freedoms to buy, sell and move their operations whenever and wherever they want around the world, unfettered by government intervention or regulations. For example, the ability of governments to use investment policy as a tool to promote social, economic and environmental objectives will be severely restricted under the MAI. Governments will have to treat multinationals 'no less favourably' than home companies. Multinationals will not have additional obligations or responsibilities to the countries they operate in under the MAI. Over 95 per cent of the largest multinational companies have their headquarters in OECD countries. Non-OECD countries will have the option of signing up to the agreement after negotiations are finalized. Under this new global constitution, multinationals will be empowered with enhanced political rights, power and security, which critics fear will effectively amount to global corporate rule.

Intellectual property rights and genetic resources from the Third World

The world's poorest nations account for around 95.7 per cent of the world's genetic resources.[1] Traditional farming practices involve farmers retaining seeds, from the harvest of one year's crop, for planting in the following year. This practice saves money on buying seed and in itself represents a continuous selection for yield and resistance to pests and diseases. However, with genetically modified seed, royalties are payable to the companies holding the patent for the seed. Under the GATT and WTO rulings, farmers have to make substantial royalty payments to multinational companies if they keep seed for replanting, even if the crop happens to be native to their particular country. In the case of hybrid crops, the seed cannot be resown successfully and must be bought each season.[25] Multinationals therefore have sound economic reasons for promoting patented transgenic seed, at the expense of other seed, for the major food crops.

The indications are that companies intend to enforce patent rulings under the WTO. This will place the onus of proof on the farmers in cases of dispute, and the resulting costs could prevent the vast majority of farmers contesting cases. Monsanto already polices its Roundup Ready™ soybean seeds in the USA, as described earlier, to check, amongst other things, that farmers do not save seeds from one harvest to plant the following year. The company uses a team of 'field monitors' to check compliance with the terms of its gene-licensing agreement with farmers.

Trade agreements and patenting legislation could restrict the traditional uses of plants in the Third World. The neem tree (*Azadirachta indica*) has been used for centuries in India as a source of insecticide, as well as having many other traditional uses. Neem trees are grown alongside fields, where the natural pesticide is made on-site by smashing the seeds, soaking them in water and scooping the emulsion from the top. It provides an economical and environmentally friendly form of crop protection. The active ingredient, azadirachtin, acts as an anti-feedant and repellent to a range of insect pests.[26] However, neither traditional methods of extraction nor modern extraction methods developed by Indian scientists are protected by national patent laws. Chemical companies in the USA have taken out a series of patents on recipes for making stable neem-based emulsions and solutions. The stability of these products allows them to be stored and marketed. For example, W.R. Grace patented a form of azadirachtin from neem seeds, which is marketed under the brand name Margosan-O.[27] The companies involved in taking out such patents maintain that their formulations are sufficiently different from the original product of nature to be patentable. Critics, however, have accused the American chemical companies of 'intellectual piracy'. Indian farmers have taken part in massive protests against the patent extensions in the GATT,[28] while Vandana Shiva, director of the Indian pressure group Research Foundation for Science and Ecology, noted that 'the novelty exists mainly in the context of the ignorance of the West'.[25]

The GATT rulings on intellectual property appear to be in conflict with resolutions passed at the Earth Summit in Rio in 1992,[29] and, more recently, at the UN Convention on Biological Diversity in Buenos Aires in 1996,[30] which state that some financial returns must be paid to countries from where genetic resources originate. A number of centres of crop diversity have been identified,[31] corresponding to where crops were first domesticated.[1] These areas include South and Central America (e.g. avocado, beans, cotton, maize, peanut, potato, red pepper, tobacco, tomato), North East Africa (e.g. banana, barley, coffee, onion, cowpea, date palm, yam, Egyptian cotton, wheat, lentil), Central Asia (e.g. almond, apple, broad bean, carrot, chickpea, onion, pea, wheat, garlic) and China (e.g. tea, millet, oat, orange, peach, rhubarb, soybean, sugarcane). All the commercial varieties of soybean grown commercially in the USA, for example, are based on a narrow range of material, representing as few as six plant introductions from the same area of China.[32] Clearly, then, most of the major food crops originated outside Western countries. A report commissioned by Christian Aid estimated that biopiracy was cheating developing countries out of US$4.5 billion a year.[33]

Large amounts of commercially valuable genetic resources from the developing world are, however, already in gene banks and botanic gardens in the developed countries. This plant material, which is outside its locations of origin, is known as *ex-situ*. Western companies are arguing that these sources of plant seed should be excluded from the United Nations Convention on Biological Diversity, which stated that companies should share their profits from plants with the countries where they were found. However, this profit-sharing provision applies only to plant specimens taken from the wild after 29 December 1993.[34]

It is estimated that as many as half the world's plants are contained in botanic gardens, most of which are in the industrialized nations. Pharmaceutical companies have approached botanic gardens in Europe and have been sold rare plants, in order to bypass the UN biodiversity convention provision by using the 1993 deadline loophole. The first reported case was a proposed contract between the US-based Phytera

Pharmaceuticals and the Palm Garden in Frankfurt, Germany.[35] In a survey conducted by RAFI, every botanic garden that replied confirmed that it had been approached by pharmaceutical companies. A major factor why pharmaceutical companies are succeeding at obtaining these resources is that many botanic gardens in the industrialized countries are now facing heavy funding cuts and are in need of the additional finance these companies can provide.[35]

However, some of the major botanic gardens are now making significant efforts to see that botanical resources are exploited to the benefit of both the developed and developing worlds. The Royal Botanical Gardens at Kew, in Britain, is building a new Millennium Seed Bank, which will extend its already extensive seed collection from around the world. The aim is to store all Britain's plant life and 10 per cent of the world's arid and semi-arid flora specimens. This £21.5 million project has been funded by Britain's National Lottery. Kew has given assurances that all requests by pharmaceutical companies will be carefully screened to ensure benefits also go to 'collaborators in the South'.[35]

Meanwhile, 'gene hunters' working for multinational companies continue to prospect for genetic resources in areas of rich biodiversity, such as tropical rainforests. The extent of this prospecting has been described by Calestous Juma, who argued that the exploitation of genetic resources has become a matter of national and regional security.[1] Multinationals employ teams of highly paid lawyers to smooth the process of patenting, while it is beyond the means of many developing world countries to establish the origin of particular genes as a native genetic resource. The provision in the Biodiversity Convention whereby developing countries should benefit from the development of genetic resources, is likely to be difficult to police. Genetic material is difficult to track and can be stored for many years. However, developing countries are now putting legal barriers in place to try to protect their biodiversity. For example, Ethiopia, an area rich in coffee and cereal species, has forbidden the export of seed from the country. Bioprospecting may soon be restricted in many developing countries due to the tightening of regulations.

Developing countries may also stand to lose substantial foreign exchange earnings as major cash crops, produced by engineering and tissue culture techniques, are grown in developing countries. Coconut oil-producing countries are threatened by crops of genetically engineered high-lauric canola, while countries producing vanilla, chocolate and pyrethrum will soon have to compete with biotechnologically produced alternatives (see Chapter 13).

Ismail Serageldin, chairman of the Consultative Group on International Agricultural Research (CGIAR), speaking at the UN World Food Summit in November 1996, feared that biotechnology patents could create a 'scientific apartheid', which locks 80 per cent of people in the developing world out of scientific advances.[36] The fear is that patent rights will continue to be extended and multinational companies will come to hold patents covering any modifications to important food crops. The flow of technology, information and genetic material to the developing world will slow down with the patenting of biotechnology. This will contrast with the situation under PBRs, where material and information was exchanged to the general benefit of Third World countries.[1] Meanwhile, the vast knowledge concerning genetic resources held by peoples in the developing countries is excluded from intellectual property rights considerations. The countries with the grip on biotechnology, it seems, are the ones who will define the future of global agriculture.

Notes

1. Juma, 1989.
2. Kung and Wu, 1993b.
3. Compare: Ducor, P., 1997, *Nature* 387: 13–14 with Helling, R. B., 1997, *Nature* 387: 546.
4. Watts, S., 1991, 'A matter of life and patents', *New Scientist*, 12 January 1991, pp. 56–61.
5. *Nature* 388: 309, 24 July 1997.
6. Stone, R., 1995, 'Sweeping patents put biotech companies on the warpath', *Science* 268: 656–8, 5 May 1995.
7. Hobbelink, 1991.
8. *New Scientist*, 26 April 1997, p. 18.
9. RAFI, 1994.
10. EPO Publication Number 0 301 749 B1. Unlike cotton, which was first developed by Agracetus using the *Agrobacterium* method, soybean was first engineered by Agracetus using particle bombardment techniques (see Chapter 2).
11. *AgBiotech: News and Information* 8 (9): 155–6, September 1996.
12. Miller, H., 'Bio-giants lobby for over-regulation' *Chemistry & Industry*, 16 December 1996.
13. Miller, 1997.
14. RAFI, 1997.
15. *Financial Times*, 7 January 1997.
16. The Uruguay Round of Multilateral Trade Negotiations opened on 20 September 1986, at the Uruguayan resort of Punta del Este. One hundred and five countries participated (including observers) and 1,500 negotiation propositions and working documents were presented. The Round closed with a ministerial conference in December 1990.
17. McCully, P., 1990, 'GATT: A brief history', *The Ecologist* 20 (6): 206, November/December 1990.
18. Goldsmith, 1990.

19. Raghavan, 1990.

20. The World Trade Organization (WTO) secretariat is located in Geneva in Switzerland. Its director-general, in 1997, was Renato Ruggiero, who has four deputies. The WTO budget is around US$83 million a year, with contributions from individual countries calculated on the basis of their share in total trade conducted. The highest WTO authority is the Ministerial Conference, which meets every two years. The first meeting was in December 1996 in Singapore. The day-to-day workings of the WTO fall to a number of subsidiary bodies, principally the General Council, which delegates responsibility to three other major bodies, the Councils for Trade in Goods, Trade in Services, and Trade-Related Aspects of Intellectual Property Rights. http://www.wto.org/

21. Erlichmann, J., 1997, 'A hard price for free trade', *Guardian*, 13 May 1997, p. 17.

22. *Nature* 387: 540, 5 June 1997.

23. Clarke, T., 1997, 'MAI-Day! The Corporate Rule Treaty'. http://www.policyalternatives.com/mai.html

24. The Organization for Economic Cooperation and Development (OECD) serves as a forum for formulating common policies among its member states. There are currently (1997) 29 member countries. All joined at the OECD's formation in 1961, unless otherwise stated: Austria, Australia (1971), Belgium, Canada, Czech Republic (1995), Denmark, Finland (1969), France, Germany, Greece, Hungary (1996), Iceland, Ireland, Italy, Japan (1964), Korea (1996), Luxembourg, Mexico (1994), New Zealand (1973), The Netherlands, Norway, Poland (1996), Portugal, Spain, Sweden, Switzerland, Turkey, United Kingdom and USA.

25. Davidmann, 1996.

26. Schmutterer and Ascher, 1984.

27. *New Scientist*, 9 October 1993, p. 7.

28. Loening, U. E., 1993, 'Freedom of farmers lost', *Guardian*, 20 November 1993.

29. The Earth Summit 1992. The United Nations Conference on Environment and Development, held in Rio de Janeiro, 3–14 June 1992. Agenda 21, Chapter 15, objective d: 'Take appropriate measures for the fair and equitable sharing of benefits derived from research and development and use of biological and genetic resources, including biotechnology, between the sources of those resources and those who use them.'

30. The UN Convention on Biological Diversity in Buenos Aires in November 1996. The main aim of this convention was to protect wild species, but the remit also included the safeguarding of protected crops in gene banks.

31. Vavilov, 1951.

32. Summerfield and Roberts, 1983.

33. Madeley, J., 1996, *Yours for Food*, Christian Aid, UK.

34. *New Scientist*, 29 June 1996, p. 7.

35. PANOS, 1995.

36. *New Scientist*, 16 November 1996, p. 6. CGIAR is based in Washington, DC at the headquarters of the World Bank. It is funded by the World Bank, aid agencies and multinational companies and oversees 18 research centres around the world.

11. Regulation of genetically modified organisms and food products

There are those who regard the application of genetic engineering in agriculture as simply an extension of existing breeding techniques, which can be encompassed within the existing regulatory framework.[1] On the other hand, there are those who regard genetic engineering as fundamentally different from previously used techniques and suggest that it should be treated as distinct, with additional risk assessment procedures and stricter regulation. Legislators have tended to favour fitting genetic engineering into existing legislation whenever possible. Genetic modifications have been treated as special cases only under certain circumstances – for example, when they result in significant alterations to food composition. The prospect of moving genes between organisms during food production, however, is one that many consumers find overwhelming. Consumers therefore need to know that adequate legislation is in place to safeguard their interests.

The regulatory frameworks in place to oversee the experimental release, development and marketing of genetically modified organisms are broadly similar in the industrialized nations.[2] The Organization for Economic Cooperation and Development (OECD), an inter-governmental forum for the harmonization of legislation, published recommendations in 1986 concerning the safety of recombinant DNA.[3] The OECD saw the establishment of a common framework as an important step to increasing the benefits of biotechnology globally, while ensuring that due regard was paid to potential concerns. The OECD recommendations were not binding on any member states, but have influenced regulations in many countries, including the USA, Germany, The Netherlands and Japan. Also in 1988, the European Commission (EC) published a framework for the regulation of biotechnology within the European Union (EU), to help member states harmonize their regulations. This led in 1990 to Directive 90/220/EEC on the Voluntary Releases of Genetically Modified Organisms into the Environment.

Many developing countries, however, lack an effective regulatory framework for genetic engineering. This could be exploited by multinationals wanting to develop or market genetically modified foods that for some reason are restricted by regulations in industrialized countries (see Chapter 14).

The regulatory framework in the USA

Regulation of genetically modified organisms in the USA is through a number of agencies working in cooperation: the United States Department of Agriculture (USDA), the Food and Drug Administration (FDA), the Environmental Protection Agency (EPA) and government departments within individual states.

The USDA is responsible for regulating transgenic plants and animals used in food production. The USDA first became a government department in 1862, created from the Office of the Commissioner of Patents, and its initial role was the distribution of plants and seeds to farmers. The USDA now regulates food plants through its Animal and Plant Health Inspection Service (APHIS) division. APHIS administers the Federal Plant Pest Act, which authorizes the regulation of interstate movement, importation and field-testing of 'organisms and products altered or produced through genetic engineering which are plant pests or which there is reason to believe are plant pests'. The application of the term 'plant pest' to a genetically modified organism means only that its 'non-pest' status has yet to be established.[4] APHIS issues permits to companies, institutions or individuals who wish to move or field-test genetically engineered plants. A detailed form (APHIS Form 2000) must be completed with details of the material to be moved or field-tested. The completed application forms are sent to the Biotechnology Permits Unit within APHIS. If portions of the application contain trade secrets or confidential information, the applicant must submit two versions, one with and one without the classified information. The latter version is sent to officials outside APHIS who need to evaluate the information, for example, departments of agriculture in the relevant states.[4]

APHIS Permits for Movement and Importation are issued or denied within 60 days of receiving an application. APHIS makes an initial risk assessment, based on information concerning the organism and its intended use, and then contacts the agriculture departments in the state or states to be involved in the movement of material. State and APHIS officials collaborate in the inspection of facilities, security and operating procedures. If a permit is granted, it is valid for one year from the date of issue.[4]

APHIS Permits for Release into the Environment[5] are usually issued or denied within 120 days. APHIS collaborates with state agriculture departments to produce a review of the proposal, including an environmental assessment. More information, however, is required on Form 2000 to apply for a permit for field-testing, or the environmental release, of a genetically modified organism. The applicant must provide details of the organism, the genes transformed and their products, the purpose of the release, the experimental design and precautions to be taken to prevent accidental release of material. Special precautions that may be necessary for movement and testing include enclosed containers for transport to field sites, field cages to prevent pollen escaping and the bagging of plants to prevent cross-pollination. If a permit is granted, APHIS personnel inspect the field site before, during and after the experimental release. Field sites are monitored for a year after experimental releases and any surviving plants are destroyed.[4]

Obtaining an APHIS field-release permit is a long process. However, in spring 1993, APHIS offered two alternatives to speed up the procedure for certain crops and situations. The first of these alternatives, the Notification Process, streamlined the permit procedure for six genetically modified crops: maize, soybean, cotton, potatoes, tomatoes and tobacco. These are crops with a history of safe release in field trials in the USA. Other crops are considered for the Notification Process on a case-by-case basis. For these crops, applicants can now just notify APHIS about proposed movement, at least ten days before transportation and field testing, and at least thirty days before any experimental releases. If APHIS considers that applications do not meet the criteria for the Notification Process, they will be considered through the regular permit process. In May 1997, APHIS announced that the Notification Process would soon be extended to all common crop species.[6] The other alternative is the Petition Process, which allows anyone to request, in writing, that a plant no longer be regulated as a 'plant pest'. If approved, the plant variety will then become exempt from the Federal Plant Pest Act and can be moved and grown without the need for APHIS permits. Petitions must include details about the genetics and origin of transformed material and should include an assessment of potential negative effects on the environment that might arise as a result of the plant's release. All petitions are printed in the Federal register. The public has 60 days to comment for or against the petition, while APHIS has up to 180 days to approve or deny the petition.[4]

If field tests are successful, a petition must be filed for USDA exemption before a genetically engineered crop can be sold commercially. This petition requires additional information to the field-test permit, including environmental product safety information.

The individual state's departments of agriculture have the authority to determine what products can be sold in the state. This can be extended to genetically modified products, as happened when Wisconsin and Minnesota declared a moratorium on recombinant BST in the early 1990s. States have the right to monitor genetically modified organisms. The way they do this varies between states. Some states, however, have more frequent applications for field releases – for example, Iowa, which has rich soils and ideal growing conditions for many of the commonly grown transgenic crops. Iowa and some other states have established formal review procedures for dealing with genetically modified plants, with advisory committees whose members are from universities and institutes within the state. The states receive permit applications from USDA–APHIS, which they review and send back.

In the USA, divisions within the USDA and FDA regulate the development of animal and human health products, including those produced using genetic manipulation. The FDA has the primary responsibility for regulating food additives and new foods, although meat and poultry are within the remit of the USDA. Recombinant BST came under FDA regulations concerning new animal drugs. The FDA has authority under the Federal Food, Drug and Cosmetics Act to remove foods it considers unsafe from the marketplace, and this act makes the producers responsible for the safety and quality of foods they market.

In May 1992, the FDA determined that foods produced from new transgenic plant varieties should be considered no differently from foods produced by traditional methods, unless special circumstances applied. The FDA cited a number of headings under which safety evaluations would be necessary, including the presence of proteins known to cause allergic reactions, changes in nutrient composition and the presence of antibiotic marker genes. FDA policy, therefore, does not require a safety evaluation review if food products are the same as those produced using non-modified crops.

The Environmental Protection Agency (EPA) regulates genetically modified organisms under the authority of two acts: the Federal Insecticide, Fungicide and Rodenticide Act, which makes the EPA responsible for regulating the distribution, sale, use and testing of pesticides, and the Federal Food, Drug and Cosmetics Act, which mandates the EPA to set tolerance levels for pesticide residues, while also monitoring any adverse effects toxins have on non-target and beneficial organisms in the field. The EPA regulates transgenic plants that contain insect toxins as plant pesticides. Therefore, transgenic plants expressing *B.t.* toxin are under the same regulations as *B.t.* toxin sprays. The EPA can also exempt pesticides from regulation if it considers them safe for human consumption. This exemption has been extended to certain transgenic plants – for example, virus-resistant plants containing genes expressing viral coat proteins, and plants containing regulator genes that enhance the activity of genes normally present in those plants. The USDA consults with the EPA when reviewing field trials of pesticidal plants.[4]

Many scientists would like to see a more rational approach to regulation in the USA. Multinationals have lobbied hard against any laws aimed specifically at the products of biotechnology, however, preferring the use of existing regulations. These commercial pressures account for the fact that a new variety of transgenic plant is considered a 'pest' by the USDA and a 'pesticide' by the EPA.[7]

The regulatory framework in the UK

In Britain, the regulatory framework has been influenced by OECD recommendations and extends the responsibility of existing organizations to include genetic engineering. The Health and Safety Executive (HSE) must be notified first, before any experimental work can be done with genetically modified organisms. The Health and Safety at Work Act of 1974 places a duty on an employer to provide and maintain a safe working environment, while also putting responsibility on employers not to expose the public to unnecessary risks. Regulations specific to genetic engineering were introduced in the Health and Safety (Genetic Manipulations) Regulations, in 1978.[2] Genetic engineering was given high-risk status so that reviews could be done before any work takes place. Proposals submitted to the HSE include details of the proposed experimental releases of genetically modified organisms, the facilities available, the monitoring to be done and the arrangements for assessing risk. The proposals are circulated to the Advisory Committee on Genetic Modification (ACGM), which evaluates each one using guidelines drawn up by the HSE and advises on acceptance or denial. Generally, the ACGM concentrates on the biological aspects of a proposal and the HSE on the physical containment aspects with on-site inspections.[8] The ACGM is a watchdog committee set up in 1984, consisting of representatives from industry, employees' unions and science specialists. It also advises the Ministry of Agriculture, Fisheries and Food (MAFF) and other government departments.

If a proposal is approved by the HSE, then further steps towards a planned release into the environment can be carried out. Initially this was under guidelines produced by the ACGM in 1986. These guidelines first required that a local risk assessment be carried out. In 1986 the ACGM guidelines were used for the first time, with the release of a genetically modified baculovirus (see Chapter 5). The Royal Commission on Environmental Pollution, building on the ACGM guidelines and the experience gained from the baculovirus releases, proposed a statuary framework for the control of releases of genetically modified organisms to the environment in 1989.[9] Releases are now covered by the 'Genetically Modified Organisms (contained use)' and the 'Genetically Modified Organisms (deliberate release)' regulations, both from 1992.

Licences for releases of genetically modified organisms are issued by the Department of the Environment's Advisory Committee on Releases to the Environment (ACRE), chaired by Professor John Beringer.[8] For pesticides notifications are also required under the Food and Environment Protection Act and under the Plant Health Act. Consent from the secretary of state for the environment[10] is needed before any actual releases of genetically modified organisms are made to the environment.

Once field trials are completed, any products need to pass pesticide, medical or food regulations. All foods are subject to the Food Safety Act, which requires that foods are fit for human consumption and not injurious to health in any way. An additional set of safeguards applies to genetically modified foods. To market novel foods, applications are made to the Advisory Committee on Novel Foods and Processes (ACNFP). This is an independent body of experts, chaired for many years by Professor Derek Burke,[11] whose brief is to advise health and agriculture ministers on any matters relating to the manufacture of novel foods or foods produced by novel processes. The ACNFP can routinely request companies to provide food composition data at regular intervals to confirm the long-term stability of genetically modified lines. The ACNFP works alongside the Food Advisory Committee (FAC) and the Committee on Toxicity of Chemicals in Food Consumer Products and the Environment (COT).

The FAC's remit is to assess the risk to humans of chemicals that may occur in or on food and to advise ministers on the labelling, composition and chemical safety of food. The FAC frequently takes advice from the COT on the toxicity of chemicals in food. The FAC then issues labelling guidelines for genetically modified foods. In 1996, it started to bring these guidelines into line with EU regulations on novel foods and food ingredients. An independent Food Standards Agency is due to be set up in 1998 in the UK to oversee food safety and policy, and will take responsibility in these areas away from MAFF.

Genetically modified foods and food ingredients approved by the

ACNFP include modified baker's yeast, enzymes for cheese production produced in transgenic yeast and bacteria, paste from transgenic tomatoes, soya from herbicide-resistant soybeans, oil from transgenic oilseed rape, maize from insect-resistant varieties, and tomatoes to be eaten fresh. However, some of these are not yet marketed in the UK because they must get clearance at European level first. Increasingly the most important decisions are being made at the European level, rather than at national level, for member states of the EU. The marketing approval for genetically modified foods in Europe is the subject of the next chapter.

Notes

1. Miller, 1994.
2. Ager, 1990.
3. OECD, 1986.
4. Webber, G. D., 1994, *Regulation of Genetically Engineered Organisms and Products*, Biotechnology Information Series: BIO-11, Office of Biotechnology, State Iowa University Extension, USA.
5. The Freedom of Information Act in the USA allows members of the public to obtain copies of APHIS permit applications for moving or field-testing genetically modified organisms, although confidential information will be deleted. Freedom of Information Act Coordinator, USDA-APHIS, LPA, PI, 4700 River Road, Riverdale, MD 20737.
6. *Trends in Biotechnology* 15: 387–9, October 1997.
7. *New Scientist*, 26 July 1997, p. 16.
8. Ager, 1988.
9. Royal Commission on Environmental Pollution, 1989.
10. The Departments of Environment and Transport were merged in May 1997 by the incoming Labour government.
11. Derek Burke retired as chairman of the ACNFP in August 1997. His replacement is Janet Bainbridge, of the University of Teeside, who took up the appointment on 1 September 1997.

12. Marketing approval for genetically modified food in Europe

Marketing approvals sought for genetically modified foods have predominantly been for ingredients of processed foods. These genetically modified ingredients have tended to become part of the general food supply rather than being discrete food items that consumers can either choose or refuse. An early example of this lack of segregation occurred in Britain and the USA, when milk from cows treated with recombinant bovine somatotropin (rBST) was pooled with the general milk supplies during the late 1980s (see Chapter 3). In this chapter, genetically modified soybeans and maize are followed through the marketing approval process in Europe, and the controversy surrounding the lack of segregation between modified and unmodified crops is examined.

Decision-making in the European Community

The European Community has four main institutions: the Commission, the Council, the European Parliament, and the European Court of Justice. The Commission consists of 17 members, who are appointed for five years. It proposes policy and legislation, executes decisions and can take legal action against member states who do not comply with Community rules. The Council discusses the Commission's proposals and makes decisions for the Community, which the Commission can amend or adopt. Membership of the Council can vary. It sometimes comprises ministers for a subject, for example agriculture, and at least twice a year the heads of government meet in a European Council. The European Parliament has 518 members (MEPs), who are directly elected for five years. Its role is to oversee and approve the work of the Commission, and it does not initiate legislation. The Court of Justice rules on the interpretation and application of Community law and its judgement is binding on member states.[1]

The Commission and Council make regulations, issue directives, take decisions, make recommendations or deliver opinions. Regulations are applicable to all member states and do not need to be approved by national parliaments. Regulations take legal precedent if there is any conflict with national law. Directives state results that must be achieved within a stated period. It is up to each member state to introduce or amend laws to bring about the desired effect. If member states fail to implement a directive the Commission may refer the matter to the Court of Justice.[1]

Decisions relating to genetically modified foods are taken under Directive 90/220/EEC on the Voluntary Releases of Genetically Modified Organisms into the Environment. This directive came into effect in 1990, although a number of amendments have been made since that time. To obtain marketing authorization within the EU under Directive 90/220/EEC, a request must be made to the member state in which the product will be first sold. A copy of this request is sent to the Commission, while advisory committees in the member state concerned, for example, the Advisory Committee on Novel Foods and Processes (ACNFP) in the UK, conduct a preliminary risk assessment of the novel food or food ingredient. The Commission forwards copies of the document submitted by the applicant, and the preliminary risk assessment, to all the member states. Any member state can make objections to the Commission within 60 days. The Commission makes the final decision as to approval or rejection.

Monsanto's Roundup Ready™ soybeans

By September 1996, a range of genetically modified food products had been approved for marketing within the USA, including produce from herbicide-resistant soybean, maize and canola; insect-resistant potato and maize; and ripening-altered tomatoes. Eighteen applications of genetic engineering in agriculture had been fully approved by the US government by 1997. A similarly broad selection of marketing approvals had occurred in Canada and Japan.[2]

The first large-scale plantings of genetically modified crops occurred in the USA in 1996; 1.2 million hectares of transgenic soybeans, cotton, maize and other crops. Soya and maize are both commodity crops, sent after harvesting to regional centres, where all the produce from farms is combined in huge bins for storage. Shipment is in large interchangeable lots, the most economical method of distribution. The seed companies, growers and shippers saw no reason why crops grown from genetically modified seeds could not also be treated in this way, and be mixed with crops grown from unmodified seed, as the products of modified seeds were deemed to be identical to those of unmodified seeds.

By September 1996, the European Commission (EC) had approved a number of genetically modified crops for the European market, under Directive 90/220/EEC. All these approvals concerned crops engineered for herbicide resistance: tobacco resistant to bromoxynil (June 1994, SEITA), oilseed rape resistant to glufosinate ammonium (February 1996, Plant Genetic Systems), chicory resistant to glufosinate ammonium (May 1996, Bejo-Zaden) and soybean resistant to glyphosate (April 1996, Monsanto).[2]

In 1996, Monsanto's Roundup Ready™ glyphosate-resistant soybean seeds accounted for around 2 per cent of the total US harvest. Soybean is an important export crop, with over 40 per cent of the total crop going to Europe. In 1997 genetically modified soya accounted for about 15 per cent of the total crop, and this figure is set to rise further. The EC approved these soybeans for the European market in 1996, despite a growing wave of concern about this transgenic crop. Critics were particularly concerned about the possible spread of herbicide resistance to other crops and weeds (see Chapter 7), possible allergic responses and the presence of antibiotic resistance genes (see Chapter 8), and the increased dependence on agrochemicals, which would increase chemical inputs to the environment and move agriculture away from sustainable systems of production (see Chapter 14).

Soybeans (*Glycine max*) were probably first domesticated in northern China around the eleventh century BC, and were not widely grown in Western agriculture until the twentieth century. This rapid expansion occurred mainly in the USA, where around 60 per cent of the total world crop was grown during the 1980s.[3] In 1994, around 51 per cent of the world crop was grown in the USA: an estimated total crop of 60 million tonnes with a farm value of around US$11.8 billion.[4] Soybean seeds are high in protein (40 per cent) and oil (20 per cent), rich in vitamins and minerals, and are an extremely versatile foodstuff, used in a large range of processed foods. The US Soyfoods Directory[5] lists the following soya products: mature whole soybeans, Edamame (Sweet Soybeans, harvested young), isolated soy protein, soy protein concentrates, textured soy protein, lecithins, Miso, Natto, soya protein and Tofu, Tofu products, frozen non-dairy desserts, Okara, soymilk, soycheese and yoghurt, soy flour (natural full-fat and deffatted), soy grits, soy meal and flakes, soynuts, soy oil, soy sauce and Tempeh. Therefore, a wide range of foods sold in supermarkets contain soya products. In processed foods, lecithin (used as an emulsifier and stabilizer), textured soya protein, soya protein concentrates, soya flour and soya oil are all common ingredients. Textured protein is usually soya protein and generic vegetable oil is often soya oil.

Processed soya can thus be found in foods as diverse as bread, biscuits, cakes, baby foods, sausages, meat substitutes, pasta, ice-cream, non-dairy ice-cream and other non-dairy desserts, chocolate and other confectionery. It is, therefore, easy to understand how 30,000 food products, and around 60 per cent of all processed foods in Britain and other industrialized countries, now potentially contain genetically modified soybeans.

The first consignment of mixed, genetically modified and unmodified, soybeans arrived from the USA in November 1996, at the Belgian port of Antwerp, seven months after the EC had approved them for the European market. Calls for the segregation of these soybeans had been repeatedly refused by Monsanto, by the exporters, including Archer Daniels Midland[6] and Cargill, and by the American Soybean Association. They all stated that in terms of safety and nutritional value genetically modified soya was equivalent to unmodified soya. This refusal to segregate soybean shipments meant that retailers, and therefore consumers, were not given the option of buying purely unmodified soybeans.

Greenpeace organized demonstrations at the European ports where cargoes of soybeans were unloaded, with protesters chaining themselves to dock railings to try and prevent unloading. The initial protest in Antwerp was stopped when Cargill obtained a court order that imposed a BF1 million fine on Greenpeace for every extra hour the protest continued.[7] Also during November 1996, at the World Food Summit in Rome, demonstrators stripped in protest against the import of modified soybeans during a press conference given by the US Agricultural Secretary Dan Glickman.[8] Meanwhile, Greenpeace and other environmental groups were also co-ordinating protests against modified soya in the USA. Activists blocked the harvest of a field of genetically engineered soya in Iowa in October 1996 and, in another incident, sprayed an entire field with pink non-toxic paint.[9]

The US-based multinational Monsanto may have underestimated public concern about genetic engineering in Europe. By choosing to ignore advice from Europe in favour of advice from economists in the USA, it alienated wholesalers, retailers and consumers who wished to receive segregated shipments. Total sales of soya to Europe were reduced by as much as 20–30 per cent as a consequence. Monsanto hopes that imports will soon return to normal, as it sees genetically modified crops soon becoming acceptable to most consumers. However, as a result of European buyers looking for alternatives to Roundup Ready™ soybean sources, Canadian soya exports, which contained no modified soya in 1996, grew by around 80,000 metric tonnes compared to the previous year.[10]

High levels of opposition to the modified soya occurred in Germany, resulting in Unilever, Nestlé and Kraft Jacobs Suchard abandoning their use of soya oil from sources containing Roundup Ready™ soybeans in processed foods in late 1996. With a significant number of consumers refusing to buy products containing genetically modified foods, it makes economic sense in certain countries to supply alternatives. Unilever Germany usually accounts for 7.5 per cent of US soybean imports into the EU, demonstrating that imports from the USA have been hit by consumer concerns.[11] Kraft Jacobs Suchard, the fourth largest food producer in Europe, stated in a fax to Greenpeace its intention to use only conventionally grown soya.[12]

In most of Europe, however, companies continued to use unsegregated supplies of soya oils in their processed food products. For example, Unilever UK continued to claim that it was not possible to avoid using genetically modified soya in products in Britain, which include Flora and Blue Band margarines, Cornetto and Solero ice-creams and Birds Eye Frozen Foods. David King, editor of *GenEthics News*, has argued that if consumer pressure had been as high in Britain as it was in Germany, a similar result could have been achieved.[13]

Switzerland, which is not a member state of the European Community, initially banned imports of genetically modified soya. In April 1997, however, it decided to reject the ban, despite protests from environmental groups and consumers, to come into line with the EC decision to approve the marketing of modified soya.[14] The decision allowed Swiss chocolate manufacturers to use lecithin products made from soya imported from the USA. However, some companies had already used lecithin from mixed shipments of modified and unmodified soya, while it was banned, and had to recall 500 tonnes of their products. Toblerone recalled 350 tonnes of its chocolate, but subsequently said that it would no longer use lecithin produced from genetically modified soya, even though it was now permitted, as 'consumers did not want it'.[14]

The EC had approved Monsanto's Roundup Ready™ soybeans in April 1996 for import and sale within the European Union. By the time the crop arrived in November 1996, however, public opinion was hardening against genetically modified food in many European countries. This meant that the EC faced more difficulty in reaching a decision for approval of Ciba-Geigy's transgenic maize, which was to arrive in European ports soon after Monsanto's soya.

Ciba-Geigy's *B.t.* maize

In September 1996, approval was pending on applications to market several more crops in Europe, including further herbicide-resistant crops – for example, oilseed rape and maize resistant to glufosinate ammonium (Plant Genetic Systems and AgrEvo) and insect-resistant maize (Pioneer Hi-Bred, Monsanto and Ciba-Geigy). Ciba-Geigy's insect resistant maize contained a *B.t.* toxin gene and also a gene conferring resistance to the herbicide glufosinate ammonium.[2] The application by Ciba-Geigy, which is now part of the giant Swiss-based multinational Novartis, was to prove the most controversial of these. Cargoes containing the modified maize were on their way to Europe as the European Commission (EC) debated its approval.

Maize (*Zea mays*) is the second most important crop in the world, in terms of commerce. The annual maize harvest in the USA in 1996 was over 560 million tonnes, of which 71 million tonnes were exported.[15] Maize exports to European Union (EU) countries are worth around US$500 million annually to the USA.[16] Modern maize varieties are hybrids, unable to survive outside cultivation. Flint or hard corn, with large floury kernels, is turned into animal feeds and used in a wide range of processed foods, including corn oil, cornflour, cornmeal, syrups, cereals, semolina, dextrose and modified starches. Maize is in fact the raw material for over 3,500 value-added products. There are many other varieties, including Sweetcorn (*Z. mays* var. *saccharata*), a variety grown as a vegetable that accumulates sugar instead of starch in its embryos, and popcorn (*Z. mays* var. *everta*), which bursts when roasted. Exports of maize to Europe from the USA, predominantly flint corn, are divided approximately 20 per cent to processed foods and 80 per cent to animal feed.

Ciba-Geigy's *B.t.* maize was engineered for resistance to the European corn borer and to the herbicide glufosinate ammonium and, like many transgenic plants, contains an antibiotic resistance gene as a selectable marker[15] (see Chapter 5). Ciba-Geigy/Novartis first applied to the French authorities for approval to place its genetically modified maize in the European market. After gaining acceptance by the French authorities, the dossier was forwarded to the EC in March 1995. It was subsequently circulated to all member states and, in March 1996, the EC proposed acceptance of the application. However, on 25 April 1996 the European Parliament voted against marketing the transgenic maize, with a number of member states either voting against the application (Austria, Denmark, Sweden and the UK) or abstaining (Germany, Greece, Italy and Luxembourg). Austria, Denmark and

Sweden objected because the EC proposal did not provide for labelling of the maize as 'genetically modified', and the UK was concerned about the risks of antibiotic resistance spreading to animals and humans.[17] However, as mentioned previously, the European Parliament has only limited influence on actual decision-making at the European level.

With the disagreement between member states on the issue, however, the Environmental Council was required to discuss the authorization to place the maize on the market. The concerns it expressed were: 1) a probable increased use of herbicides; 2) the continuous expression of *B.t.* toxins leading to possible resistance in insects, which may make *B.t.* sprays used by organic growers and in biological control programmes less effective; 3) the use of the antibiotic ampicillin as a marker possibly transferring antibiotic resistance to micro-organisms in the gut; and 4) the possible allergenic effects of the new enzymes expressed in plants used as food. These concerns corresponded to those raised by environmental and consumer groups, although these groups were additionally concerned about the lack of segregation and labelling.

The European Commission then entered a period of protracted debate over the Ciba-Geigy/Novartis maize. It decided to leave its final decision until three specialist committees (the Food, Animal Nutrition and Pesticides Committees) had reported with their recommendations. These committees got behind schedule, and reports were continually being delayed through November and December, as shipments of modified maize arrived in European ports. Pressure was growing for a quick resolution, as cargoes containing maize were held at the ports. On 10 December, the EC revised Directive 90/220/EEC to simplify marketing procedures.[18] Then on 18 December 1996, the EC approved the modified maize for sale in Europe, after the three scientific committees had agreed that there was no risk to humans or animals from consuming it.[19] A source close to the Environment Commissioner said: 'We have been waiting for these opinions for six months and it was difficult for us not to accept them today.'[20] Consumer and environmental groups reacted strongly against the decision, seeing it as favouring political and commercial interests over public and environmental safety. Greenpeace called it 'one of the most irresponsible decisions the Commission has taken'.[20]

The decision meant that thousands of tons of maize, held in European ports, could enter the market, while food products containing the maize as an ingredient would not require any special labelling. Critics argued that the EC had caved in under intense pressure from the USA and multinational companies. The US secretary of agriculture, for instance, sent a delegation to Brussels to lobby against any restrictions on marketing, arguing that objections were based on 'unsound science' and would constitute a trade barrier. A ban on the maize could therefore have triggered an immediate trade war, as the US considered a ban illegal under the free trade rulings of the World Trade Organization (WTO). To defend any ban, it would have been necessary to demonstrate that the crop failed to meet agreed standards, or to present firm scientific evidence of health or other risks. The USA is likely to use its powers under the WTO if it feels any of its transgenic crop exports are being unfairly treated in Europe.

The EC admitted, in December 1996, that illegal imports of genetically modified maize from the USA had occurred since the first of October. According to import certificates, 4,000 to 5,000 tonnes of the crop, which included a proportion of maize grown from the modified Ciba-Geigy/Novartis seed, had been arriving weekly through the ports of Antwerp, Rotterdam, Lisbon and Barcelona.[21] Imports were in effect illegal until the decision on genetically modified maize had been made by the EC. The Commission appeared to be powerless to prevent the imports, relying on member states to carry out adequate inspections to ensure that Community legislation was applied. Critics pointed out that, because most border controls between member states had been abolished within the single market, maize imported by one country could move freely to any other member country. The Commission saw no point in taking action so close to a decision being made on maize imports. Member states, many with major reservations about the transgenic maize, were left virtually powerless by the Commission's decision. There were few legal options available to prevent the maize imports, apart from threatening to bring the Commission before the European Court of Justice for violating Directive 90/220/EEC, a threat that would have put pressure on the Commission to withdraw the approval.[17] This did not occur, but some member states did use a safeguard clause to ban imports temporarily.

In order to have stopped illegal imports during 1996 it would have been necessary to prove that modified maize was in a particular cargo. Testing shipments for genetically modified content has been likened to finding a needle in a haystack. Genetically modified maize formed less than 1 per cent of the crop and scientific testing would have been required to distinguish it from non-modified maize.[22] Tests are now becoming available, however, for identifying modified produce within shipments. For example, Genetic ID, a company in Iowa, uses equipment that scans the DNA structure of crop samples and identifies any altered gene sequences. The equipment can detect one genetically modified maize kernel in 10,000 non-modified kernels. The company is working with major European food producers who are considering the test for monitoring shipments of soya and maize in the future, when genetically modified produce will form a much higher proportion of the total crop. One company, Central Soya, was supplying certifiable non-Roundup Ready soybeans from late 1996. They were able to do this by using soya from only Canadian sources, where modified beans were not grown at that time. Guaranteed non-modified supplies from countries in South American were also available.

In an additional twist, in January 1997 a US academic raised fears over the reliability of testing of genetically modified maize. Margaret Mellon claimed that the maize had been initially cleared in the US by the Food and Drug Administration (FDA) and other advisory committees without them knowing about the selectable marker gene: that is, the decision was made on the knowledge of herbicide and insect resistance genes alone.[23] It was the presence of this marker gene, conferring antibiotic resistance, that was to cause European advisory scientists most concern.

On 6 February 1997, Austria became the first European country to make it illegal to import and market Ciba-Geigy/Novartis' transgenic maize on its territory, despite the EC marketing approval of 18 December 1996. This involved using a safeguard clause (Article 16) in Directive 90/220/EEC that allows member states to ban, for three months, the sale of modified products that already have EC approval, if they believe them to be a risk to the environment or to health. Austrian Health Minister Christa Krammer justified the decision by referring to continuing doubts within her ministry concerning the effects on health of the transgenic maize.[24] Luxembourg followed, on 7 February, with its own ban on the same safety grounds. Both Austria and Luxembourg were concerned about the selectable antibiotic marker gene and the risk of resistance developing in gut bacteria to ampicillin and other penicillin antibiotics. The European Commission had three months to decide if these bans constituted an unnecessary obstruction to the free movement of goods within the European market or whether risks to the environment or health warranted an extension of the ban throughout Europe.

France, which was active in getting the maize accepted on the EU market, said that it would authorize the marketing of genetically modified maize, but only if it was appropriately labelled. On 13 February 1997, it further announced that the cultivation of the modified maize was prohibited within France. This prompted the head of the nation's Biomolecular Engineering Commission, Alex Kahn, to quit. He claimed that the government's 'incoherent policy' of saying the maize was safe to eat, but not to grow, gave him no choice but resign. Italy became the second EU country to ban the cultivation of the modified maize on 4 March. France and Italy grow two-thirds of the EU's annual maize crop of 33 million tonnes.[25]

The different policies adopted by member states towards genetically modified crops opened serious splits within the European single market. By the spring of 1997, however, 13 out of 15 EU member states had doubts about the authorization of the transgenic maize. The EC said that further amendments to Directive 90/220/EEC would be prepared in an attempt to resolve the conflict. Member states were under pressure to take some sort of action, however, even if only of symbol value, as public opinion appeared to swing against the import and marketing of unlabelled foods from transgenic crops.

On 8 April 1997, the European Parliament condemned the European Commission for authorizing the genetically modified maize imports. MEPs voted resoundingly (407 votes for and 2 against) for a resolution that accused the Commission of a 'lack of responsibility' in approving the Ciba-Geigy/Novartis maize, despite the earlier votes against its approval by member states and the European Parliament. It further claimed that 'trade considerations have obviously dominated the decision-making process so far'. The resolution stated that 'serious doubts remain as to the safety of genetically modified maize and the risks of transmission to human beings of a marker gene resistant to antibiotics'. It called for the publication of the complete findings of the three scientific committees, on whose advice the Commission had authorized the importing of the maize, and for a revision of procedures for the authorization of marketing of genetically modified food products to 'correctly reflect the democratically expressed opinions of the member states and the European Parliament'.[22] During 1997, member states, including Austria, Luxembourg and Italy, also became increasingly concerned about the possibility of *B.t.* genes in transgenic crops causing insects to develop resistance to *B.t.* sprays, an important component of pest control on organic farms.

In a MORI poll, commissioned by Greenpeace and conducted in five European countries,[26] 59 per cent of people were opposed to the development and introduction of genetically modified foods, with only 22 per cent being in favour, and 67 per cent saying they would not be happy eating such foods.[27] A Greenpeace spokesperson, commenting on the MORI poll, said: 'Legislators are obliged to respect such a clear preference by European consumers.' Different attitudes to genetic engineering and food production, however, exist around Europe. In the

MORI poll, Swedes were most strongly opposed (76 per cent against) and Britons the least concerned (53 per cent).[27] Germany and Austria, two countries not included in the MORI poll, have the highest levels of opposition to genetically modified foods in Europe.

In early 1997, Germany witnessed major public demonstrations against nuclear power and the cloning of animals, as well as against genetically modified foods. These environmental issues are a reflection of a strong Green political movement within Germany, which is particularly concerned about the misuse of genetics in the light of discredited eugenics policies in the country's past. A poll in Germany, conducted by the GfK Institute, showed that 80 per cent of the population did not want to eat genetically modified foods.[10] Consumer pressure had already caused food-processing companies in Germany to abandon their use of genetically modified soya, while wholesalers and retailers in Germany have been active in locating sources of non-modified maize for the home market.

Polls in Austria during the early part of 1997 revealed that around 85 per cent to 90 per cent of the population supported a referendum on the issue of genetically modified food. Two-thirds of supermarkets in the country pledged not to stock genetically modified foods, and the two most popular newspapers in Austria campaigned daily in support of bans on genetically modified foods. The strong Green movement in Austria had previously succeeded in gaining a referendum in 1979, which had led to the banning of nuclear power. A referendum or *volksbegehren*, initiated by a coalition of environmental groups and organized by the Austrian minister of internal affairs, was held over the week ending 15 April 1997, and resulted in 1.2 million people, one-fifth of the Austrian electorate, signing a petition against the use of genetic engineering in food production.[28]

The result of the referendum is not legally binding on the Austrian government, but put considerable pressure on them to act on the issue. The government at the time, a coalition between the centre-left Social Democrats and the centre-right Austrian Popular Party, discussed the specific demands arising from the referendum. These demands included a ban on the production of genetically modified food in Austria, a moratorium on field-testing of transgenic crops and a ban on imports of transgenic soya. By accepting a number of the proposals in the referendum, the government knew it could influence future EU policy. Austria was already in potential breach of EU legislation because of its temporary ban of 6 February on the import of Ciba-Geigy/Novartis' *B.t.* maize. The government responded to the referendum by refusing to back down on this ban, and told the EC it was willing to go to the

European Court of Justice if the EC ruled its ban illegal. Switzerland will hold a similar referendum on genetic engineering in 1998.

The EC was due to deliver its decision on Austria's banning of Ciba-Geigy/Novartis' *B.t.* maize on 14 May 1997, when the three-month ban expired. A European-level committee had considered the issue, but said it had not been able to conclude its deliberations on time. The EC was in a difficult position, because if it extended the ban on the maize the USA would consider it a violation of free trade rulings. The EC could either rule the ban illegal, further displeasing the governments of several member states, or extend the ban throughout Europe, in which case the WTO would probably, under intense pressure from US-based multinationals, rule the ban illegal. Eventually, in September 1997, the EC requested that Austria, Luxembourg and Italy drop their unilateral bans on Ciba-Geigy/Novartis' maize. At the time of writing, Austria is seeking a judgment from the European Court of Justice. Similar conflicts over marketing approvals are likely to be played out with other transgenic crops.

In September 1997, Norway announced plans to ban six genetically modified products that had been authorized by the EC, including Ciba-Geigy/Novartis' maize. Norway is not a full member of the EU, but is a part of the European Economic Area, set up in 1992 to extend the EU's single market. Nevertheless, the EU threatened to take retaliatory action against Norwegian products if the ban was implemented.

A new wave of crops

Further imports of mixed shipments of genetically modified and unmodified crops to Europe followed in 1997, for crops that had received EC approval. These included transgenic varieties of maize from Monsanto, AgrEvo, Pioneer Hi-Bred and Northrup King. These crops were engineered for insect and herbicide resistance characteristics. Oil from herbicide-resistant canola was imported from Canada for the first time in 1997, by AgrEvo, and will be used in margarine and other processed foods. Some of these transgenic crops are now also being grown commercially within Europe.

The multinationals during 1997 continued to justify the lack of segregation because of the similarity between foods produced from modified and unmodified ingredients, while continuing to resist calls for labelling (see Chapter 13). European food industry associations, meanwhile, continued to disagree among themselves on whether genetically modified crops could feasibly be segregated from non-modified varieties.[16] The Grain and Feed Trade Association (Gafta), which represents world bulk traders, stated in March 1997 that segregation was 'no longer a viable option' for that year or any future year. The European Retailing Association 'Eurocommerce', on the other hand, stated that segregation would be possible if determined by market forces. European retailers, influenced by their consumers, have called for segregation and labelling whenever possible. However, Gafta responded by saying that US farmers would need 'substantial' incentives to introduce segregation.

The agrochemical and biotechnology industries in Europe, meanwhile, fear they will be left behind if further obstacles are put in the way of marketing genetically modified crops. Multinationals have exploited this 'fear of Europe being left behind' argument to exert political pressure on decision-makers. In early 1997, eight genetically modified food product releases had been approved in the EU, compared to two dozen in the USA.[29]

The approval of further herbicide- or insect-resistant crops was pending under Directive 90/220/EEC during 1997, including canola (spring oilseed rape) resistant to glufosinate ammonium (Plant Genetic Systems), and various insect-resistant maize varieties (Northrup King and Pioneer Hi-Bred). The EC approved two varieties of Plant Genetic Systems' canola in June 1997 for use in animal feed and for further transgenic seed development. The company, which is now part of AgrEvo, had given assurances that seed bags containing transgenic seed sold to farmers would be labelled as such. These transgenic seeds were available to EU farmers from July 1997. However, Austria has stated that it intends to ban imports of Plant Genetic Systems' modified oilseed rape.

A range of other genetically modified crops and foods were also pending approval by the EC by 1998. These included transgenic tomatoes for eating fresh. Calgene's Flavr Savr™ tomatoes are marketed in the USA and have already been approved by the UK government, following advice from the ACNFP. They may become the first fresh transgenic fruit or vegetable to gain marketing approval in Europe. This is likely, therefore, to be a key decision.

Notes

1. Goodman, 1993.
2. *Europe Environment* 490: 19–22, 17 December 1996. By September 1996, the following transgenic crops had been given approval for use in foodstuffs in Japan: glyphosate-resistant soybeans and canola (Monsanto); various glufosinate ammonium-resistant canola varieties (AgrEvo and Plant Genetic Systems); and insect-resistant maize varieties (Ciba-Geigy and Northrup King). http://ss.s.affrc.jp/docs/sentan/eguide/commerc.htm

3. Summerfield and Roberts, 1983.
4. RAFI, 1994.
5. *US Soyfoods Directory* 1996. Indiana Soybean Development Council, USA.
6. Archer Daniels Midland (ADM) is an exporter of modified soya and maize. The company also manufactures sweeteners from maize, which, along with ethanol production, accounts for half its annual profits of around US$700 million. ADM has vigorously lobbied the US government to promote its sugar substitutes at the expense of sugar imports from developing countries. See Chapter 13.
7. *Nature* 384: 102, 14 November 1996.
8. *Guardian*, 14 November 1996, p. 6.
9. *AgBiotech: News and Information* 8: 197N, December 1996. Greenpeace resumed its protests against genetically modified soybean and maize imports during October and November 1997, taking direct action to impede the unloading of ships and block the movement of lorries. It established a Genetics Hazards Patrol to monitor shipments arriving from the USA at European ports and to track the movement of imports along European transport routes (see, for example, http://www.greenpeace.org/~comms/97/geneng/press/november11.html). Meanwhile, in Ireland, the country's first genetically modified crop was destroyed by a group of environmental activists calling themselves the Gaelic Earth Liberation Front. Monsanto was growing the crop of Roundup Ready™ sugar beet as part of a three-year project approved, in the face of widespread opposition, by Ireland's Environmental Protection Agency. *Nature* 389: 534, 9 October 1997.
10. *Biotech Reporter*, January 1997. In contrast, Canada stood to lose 400,000 tonnes of canola sales to Europe in 1997 because of the presence of seed from transgenic varieties in unsegregated shipments. *Food Ingredients and Analysis*, September 1997, p. 3.
11. *Nature* 384: 301, 28 November 1996.
12. Greenpeace Press Release. 'Retailers and Food Manufacturers: Who is doing What?' 22 November 1996. http://www.greenpeace.org.uk/science/ge/index.html
13. King, D., 'New beans means profits', *Guardian*, 11 December 1996, p. 5.
14. *Nature* 386: 479, 3 April 1997.
15. *Documentation on Bt-maize from Ciba Seeds*, 1996, Ciba Seeds Communication, Basle, Switzerland.
16. *ENDS Report* 264, January 1997.
17. Greenpeace Press Release. 'Background on Genetically Modified Maize'. December 1996. http://www.greenpeace.org.uk/science/ge/index.html Article 4, Paragraph 1, of Directive 90/220/EEC states: 'Member states shall ensure that all appropriate measures are taken to avoid adverse effects on human health and the environment which might arise from deliberate release or placing on the market of genetically modified organisms'.
18. Report COM(96)630 adopted as revision to Directive 90/220/EEC on the voluntary release of genetically modified organisms.
19. Report COM(96)206 giving approval for marketing of Ciba-Geigy maize in compliance with Directive 90/220/EEC on the voluntary release of genetically modified organisms.
20. *Europe Environment* 491: 9–11, 14 January 1997.
21. *Nature* 384: 503, 12 December 1996.
22. *Guardian*, 7 December 1996, p. 10.
23. *Farmers Weekly*, 10 January 1997, p. 10.
24. *Science* 275: 1063, 21 February 1997.

25. *ENDS Report* 266, March 1997.

26. MORI poll: 4,840 people interviewed in Denmark, France, The Netherlands, the United Kingdom and Sweden, 11–20 December 1996.

27. Fifty-nine per cent of Europeans surveyed in MORI poll reject transgenic food. *Europe Environment* 491, 14 January 1997.

28. *Europe Environment* 498: 8–9, 22 April 1997.

29. *Farmers Weekly*, 1 November 1996, p. 20.

13. The consuming question of labelling

As consumers have become aware of the extent to which genetically modified ingredients are used in processed foods, there have been growing calls for these foods to be labelled. The food industry has resisted the labelling of most genetically modified foods, on the basis that these foods are equivalent to foods produced with non-modified ingredients. In this chapter, the arguments for and against the mandatory labelling of all genetically modified foods are examined, and the development of labelling legislation in Europe is described.

Lessons from irradiated foods

Labelling can be bad for business, as the food industry learned from its experience with food irradiation. This technique was developed to increase the shelf-life of fruit and vegetables. Irradiation involved bombarding food with gamma rays, which stemmed the rotting process, inhibited sprouting and killed bacterial contaminants. As with transgenic fruits produced for longer shelf-life, irradiation was claimed to leave food as safe and nutritious as untreated food, although critics argued that irradiation could be used to disguise food that would otherwise be unfit for sale.[1]

Irradiation techniques were refined during the 1970s, when recommended irradiation levels were set, based on data from an extensive research programme. However, consumer groups mounted effective protests against the use of irradiation as a food preservation method. The irradiation critic Richard Piccioni linked the technology to the nuclear industry in the USA, where potentially hazardous by-products of nuclear weapons manufacture (caesium-137 and cobalt-60) were being proposed as gamma ray sources for food irradiation. He warned of the dangers of contamination by carcinogens.[2] In the UK, concern among consumer groups grew during the late 1980s, fuelled by reports that high levels of irradiation could destroy vitamins in food.[1] Despite food industry reassurances, for example, that cooking destroys more vitamins than irradiation, consumer pressure meant that supermarkets introduced voluntary labelling on all irradiated foods. This was followed by mandatory labelling. Given the choice, consumers preferred not to buy products that had been treated with ionizing radiation. Because of poor sales, supermarkets stopped stocking irradiated foods.

The case against mandatory labelling: the food is no different

The food industry fears a similar consumer response to food labelled as being genetically engineered, and so is against mandatory labelling. A distinction needs to be made between those foods that are themselves, or contain, genetically modified organisms, and those foods that are produced by genetic engineering processes. Although fruits and vegetables modified for longer shelf-life, taste or composition are clearly genetically modified organisms, and are likely to be labelled as such, most genetically engineered food products are processed foods. The foreign genes themselves are often destroyed during processing, and the end-products are in many cases identical to products made from unmodified material.

The first genetically modified foods sold in the UK were tomato purée and vegetarian cheese. The Sainsbury and Safeway supermarket chains stocked the tomato purée, which was not legally required to be labelled. However, both these supermarkets decided to label the purée voluntarily.[3] Vegetarian cheese, made using genetically modified chymosin, is not required to be labelled any differently from other cheese in either the UK or the USA. In the UK, the Co-op supermarket chain stocked this product and labelled it voluntarily.

In many cases there is no scientific evidence to suggest that the methods of production using genetic engineering *per se* alter food composition in a meaningful or uniform manner. The food industry understandably wants labelling to be on a strictly logical and scientific basis. Monsanto has said that if products made from modified soya oil could be shown to be different from products made from unmodified sources, then it would support their labelling. In this, it is following recommendations made by Codex, an international body formed to oversee food standards,[4] that mandatory labelling is necessary only when material differences can be shown in food due to the genetic modification process. The genetic change, in the case of Monsanto's modified soya, was for a protein that modifies the enzyme pathway blocked by a herbicide. The foreign gene should not find its way into oil products of soya, although it may be present in other processed foods made from soya. Monsanto further supported its case by publishing data showing no biochemical differences between the chemical composition of modified and unmodified soybeans.[5]

The food industry argues that because traditional methods of genetic change, such as cross-hybridization of crop varieties, are not required to be stated on labels, then genetic engineering techniques should not have to be stated either, as the final food products are no different. Generally, the information on food labels pertains to the composition and attributes of the food and not to the details of the manufacturing process. Labelling all the foods that use genetic engineering somewhere in their production would send the consumer signals that the foods were in some way unsafe, according to the food industry, when this is not the case. Labels could, therefore, unjustly stigmatize genetically modified foods. The differential labelling of equivalent products from different countries may also fall foul of international free trade agreements, if one country could show that its products were being unfairly marketed. Consumers are often not well informed about new technology in food production and may feel uneasy about alterations in the traditional food supply. A process of education, says the food industry, will reassure the public about this new food production technology.

As described earlier, commodity crops, like soya and maize, are grown over large areas in big industrial farming operations. The produce from entire regions is pooled and sold in bulk and therefore modified and unmodified crop is mixed together, making labelling further down the distribution line difficult. For mandatory labelling to be most effective, genetically modified foods would need to be segregated at an early stage. This is not considered desirable by the food industry, particularly as it considers the end food products to be identical. The commodity crop producers claim that segregation would require the development of new food distribution systems and cause major disruption to existing national and global distribution systems. In the case of some commodity crops, the cost of segregation may exceed the value of the product. Therefore the very attributes, such as insect resistance or herbicide resistance, that allow farmers to produce food more economically would effectively be barred from use.

Mandatory labelling is also resisted for political reasons. A number of UK politicians called for compulsory labelling in 1996, but the House of Commons Select Committee on Science and Technology recommended that no compulsory food labelling should occur. Both the Polkinghorne Committee of 1993[6] and, more recently, the Food Advisory Committee (FAC), which is responsible for labelling decisions in the UK, agreed that it would be unrealistic to label every food produced with the aid of genetic engineering.

Mandatory labelling might also jeopardize the continued development of genetically modified foods because of initial consumer resistance. This would be bad for the industry as a whole, at a time when many industrialized countries wish to encourage biotechnological development, and could lead to the abandonment of agricultural projects that may yield important benefits in the future.

The case for mandatory labelling: the consumer's right to choose

Whatever the biochemical changes or calculated risk, labelling represents the right of consumers to know what is in their food, and the right to choose what foods to buy and eat. Consumers groups have been worried by this denial of the consumer's fundamental right to choose. Many consumers and environmental groups also see fundamental differences between food produced using genetic engineering and food produced using other techniques of genetic improvement, such as traditional plant-breeding methods. They are concerned that unpredictable changes in food composition may occur and make these foods unsafe in some way. Therefore, they argue, all genetically modified foods should be labelled so that the public can make informed purchasing decisions.

Consumers may have ethical or moral objections, for whatever reason, to genetic engineering *per se*. However, the food industry's insistence that labelling be based on scientific principles will limit the freedom of choice of these consumers. Although it can be shown scientifically that foods made from modified crops are identical to foods made from unmodified crops, consumers might want to avoid these foods simply because of their method of manufacture.

The assertion that foods made using modified and non-modified ingredients are equal is based on the principle of substantial equivalence. This involves quantifying a number of selected characteristics of the food, and if these are found to be equivalent then the foods are assumed to be equivalent in all other characteristics. Critics of this approach stress that it focuses on potential risks that can be anticipated on the basis of known characteristics, while ignoring the unexpected risks that may arise when organisms are modified using genetic engineering.[7]

Genetically modified foods may also contain the antibiotic resistance genes, used as selectable markers, which, although not affecting the nutritional composition of foods, may be of concern. Consumer groups have suggested that the presence of these, and of other types of marker genes, warrant mandatory labelling of a food product, irrespective of the presence of other genes. In the case of Ciba-Geigy/Novartis' maize, an antibiotic marker gene caused as much concern as genes expressing insect-specific toxin and an enzyme that prevented herbicide detoxification, because of its potential risk to human health.

Calls for clear and meaningful labelling of genetically modified foods gathered support through the 1990s. The mixed consignments of modified and unmodified soybeans in 1996 gave retailers and consumers no choice when buying processed foods, which might or might not have contained genetically modified soybeans. By December 1996, more than three hundred consumer, health and agricultural organizations had launched a worldwide boycott of genetically modified soya and maize. Participating organizations urged consumers to boycott targeted products, including Nestlé Crunch, Similac Infant Formula, McDonalds' French Fries, Kraft Salad Dressing, Quaker Oats Corn Meal and Coca-Cola.[8] In the USA, the Foundation on Economic Trends, and its director Jeremy Rifkind, have been particularly active in mobilizing opposition to genetically modified foods. Their attitude on labelling regulations is: 'If food producers are so proud of these "Brave New World" products, why are they so afraid to label them?'[9] One vocal celebrity group of protesters who have been mobilized are chefs. In the USA, 1,500 chefs joined the Pure Food Campaign and displayed 'We do not serve genetically engineered foods' stickers on their menus.[9] In Britain, top chefs also expressed their concern about the lack of labelling on foods they supply to their customers, which might contain genetically modified produce.[10]

Europe decides

On 12 March 1996, the European Parliament, the parliamentary assembly of the European Union (EU), adopted a resolution calling for all genetically modified products to be labelled as such and sold separately from non-modified products. However, the Parliament, whose role is essentially advisory, was still some way from having its demands incorporated into EU legislation. Green groups were rightly concerned when the European Commission (EC) sought to bypass five (out of six) key amendments passed by the European Parliament in legislation on the labelling of genetically modified food. The European Parliament's stance was at odds with the EC's desire to avoid compulsory labelling. As with the decision to approve genetically modified foods for the

European market, a major factor in the EC's reluctance to accept labelling was that it might trigger a trade war with the USA.

American-based multinationals had argued that labelling would constitute a disguised barrier to trade, favouring unlabelled non-modified foods in Europe over equivalent but labelled foods from the USA. The modified food contains the same ingredients, but just happens to be produced using genetic engineering. Trade agreements do not consider the methods of production to be significant. Free trade rights are upheld by the World Trade Organization (WTO), which has powers to impose sanctions against countries that ignore its rulings. The WTO rules require the importing country to show evidence that a product is harmful rather than the exporting country to prove that it is safe. The USA has defended its trade interests using trade agreements on a number of previous occasions. For example, when Canada acted to ban irradiated foods because of consumer concerns, the ban was deemed illegal under the terms of the Canada–US Free Trade Agreement.[11] Companies in the USA used the WTO to overturn an EU ban on products from BST-treated cattle, as noted previously. Meanwhile, a proposed European ban on fur from animals caught in leg-hold traps has repeatedly failed due to claims by the USA, Canada and Russia that it would infringe free-trade rights. The EC has also shelved a Cosmetics Directive, designed to ban the sale of products tested on animals, because of its possible violation of free trade agreements.[12]

By the end of 1996, Germany, Austria, Denmark and Sweden supported full labelling of all genetically modified foods.[13] Around this time, the Environmental Council of the European Parliament expressed concern about the large number of authorizations being made by the EC under Directive 90/220/EEC, in the months before a proposed new regulation became law. The Environmental Council claimed that this was prejudicing the question of labelling, as the new regulation had stricter procedures for labelling.[14]

In December 1996, after a protracted debate, the EU agreed on this new Novel Food and Food Ingredient Regulation. The regulation represented a compromise on labelling, and did not require labels on all food produced from genetically modified organisms. Labelling, under the new legislation, would be required for novel foods only if they contained viable ('live') genetically modified organisms, had modified ingredients that were no longer equivalent to existing ingredients, or contained materials that were not present in the original foodstuffs, or substances that may give rise to ethical concerns, such as animal genes.[15] Applicants would have to submit a label for consideration for genetically modified foods in the above categories. The Novel Food and Food

Ingredient Regulation was approved by the European Parliament in Strasbourg on 16 January 1997 and came into force on 15 May 1997, as Regulation 258/97/EC. The regulation applied only to foodstuffs or food ingredients put onto the market after 15 May, and did not apply to products already approved or to the eleven products in the approval's pipeline. Therefore, the transgenic maize and soya, which were already approved for market, did not require labelling. The European Parliament, however, called for the EC ruling to be applied retrospectively so that it covered Ciba-Geigy/Novartis' modified maize.[16]

The Novel Food and Food Ingredient Regulation was attacked for being too vague, subject to interpretation and too broad by a range of environmental groups, including Greenpeace, who argued that it provided loopholes to those wanting to avoid labelling foods containing genetically modified ingredients.[15] Foods did not require labels if 'substantially equivalent' to existing products. Therefore processed foods, such as those containing genetically modified soybeans, did not require labelling, as they were considered 'dead' or not 'substantially different' from equivalent unengineered foods. Other processed foods, such as tomato paste and vegetarian cheese, also did not require labelling under the regulation. The regulation applied, however, to all fresh fruit and vegetables. For example, Calgene's Flavr Savr™ tomatoes, approved by the UK for eating fresh, and during 1997 being considered by the EC under Directive 90/220/EEC, would require labelling. Only between 5 per cent and 10 per cent of all foods produced using genetic engineering therefore required labelling, according to the Genetics Forum.[17] Greenpeace also criticized the regulation for not laying down rules about positioning, wording or size of labels.[15] This might lead to labels saying 'produced with the latest technology' or using other vague wording. In addition, the regulation did not address the complexities of the food supply chain and did not tackle the important issue of segregation.

The new regulation was similar to the food labelling laws in the USA at that time, where no labelling of food made using any genetic engineering or biotechnological processes is required if the composition of the food is substantially equivalent to that produced using traditional methods. However, consumer and political pressure was growing for stricter labelling legislation in Europe.

If processed food products manufactured from mixed imports of modified and unmodified crops were to be labelled as possibly containing genetically modified material, then such labels would be 'meaningless', according to the European Bureau of Consumer Unions (BEUC). Such labels would be widespread – for example, given the ubiquitous nature of soya products in processed foods – and would give confusing information to the consumer. BEUC called on retailers and the food industry to exert pressure on suppliers to segregate shipments, so that effective labelling could be introduced. The Novel Food and Food Ingredient Regulation, however, was an initial attempt to let consumers make some sense of the confusion caused by the lack of segregation, by imposing labels on those products where genetic engineering has made the greatest change to food composition.

Retailers felt relatively powerless to control what they bought, in the way of soya and maize, because of the mixed shipments in 1996. For example, a technical manager at the Iceland chain, in the UK, expressed anger and frustration and claimed that it was 'irresponsible of Monsanto to allow products into the marketplace which cannot be adequately labelled'.[18] In late 1996, the Iceland and Co-op retail groups were supporting consumer calls for full labelling of genetically modified foods. Most UK supermarkets appear not to be opposed to genetically modified food in principle, but say they would like to offer customers a choice by providing guaranteed non-genetically modified produce. Tesco, Iceland and Somerfield opted to use only non-modified soya in their own-brand products and some US distributors have made agreements to provide sources of non-modified soybeans from 1997.[18] Therefore, in practice, mandatory labelling would not now result in the widespread labelling of all processed foods, because retailers are seeking out non-modified foods and ingredients. Canada's exports of soya, for example, have risen as the demand for sources of non-modified soybeans increased.

Supermarkets in other parts of Europe are also responding to increased public concerns. For example, the Swiss supermarket chain Migros banned products likely to contain modified soya from its shelves in early 1997. Only a relatively small proportion of the soya and maize crop was modified in 1996, but a higher proportion of the 1997 harvest was grown from modified seed and this proportion will increase in coming years, leading further retailers to decide that the labelling issue has now become of sufficient importance to need addressing.

The food industry in The Netherlands announced in December 1996 that it would mention on the labels of their food products if any genetically modified soya ingredients had been used. The labels appeared in shops from April 1997 and represented the first such move in Europe. This also opened up a potential gap in the European single market, as other European countries could regard the Dutch move as a barrier to trade.[19] Soon after the Dutch decision, France and Denmark also announced that they would require labelling on processed foods made with genetically engineered soya and maize, while in March 1997 Austria used an Environmental Council meeting to call for an extension of labelling requirements. In response to these initiatives, the EC stated that member states could impose national labelling laws for genetically modified foods.

A change in EC attitude towards a more consumer-orientated view was signalled by EU Agriculture Commissioner Franz Fischler in April 1997, via his homepage on the World Wide Web.[20] Here, he expressed his personal view that labelling should be applied regardless of whether or not the foods are different from traditional products, while segregation and labelling should be carried out all along the production and distribution chain, from the farm to the retailer.

The food industry has consistently maintained that segregation is not a viable option, but consumer groups and retail organizations in Europe, including the British Retail Consortium, believe that segregation ought to be feasible. The industry's claim was undermined by reports that genetically modified maize had been segregated informally by farmers and grain dealers in the USA. In 1997, Ciba-Geigy bought back a substantial proportion of its *B.t.* maize to sell as seed for the 1998 growing season, while Mycogen promised to buy back any of its *B.t.* maize (NatureGard™) that farmers were unable to sell. Segregation would be necessary in these cases. Meanwhile, some growers and dealers segregated their maize, because they were aware of the controversy in Europe and did not want a small proportion of modified maize affecting shipments.[7]

The EC hardened its attitude on labelling during June and July 1997, in response to mounting pressure from member states and consumer groups, when it adopted new rules requiring the mandatory labelling of all genetically modified foods. It agreed on 23 July that manufacturers should label all products, including processed foods, to indicate whether they 'do contain' or 'may contain' genetically modified ingredients. The proportion of genetically modified material in products could also be stated. This meant that from 31 July 1997 companies had to submit a proposed label as part of the marketing approval process. These new labelling guidelines were incorporated into Directive 90/220/EEC, as part of a major revision of this directive in late 1997. The rules did not, however, go as far as requiring segregation.[21]

The EC had hoped that the guidelines would strike a blow against segregation, as mixing modified products with non-modified products would no longer exempt foods from being labelled. To avoid widespread use of the 'may contain' label, the 'contains' label will be used whenever there is any evidence that products contain genetically modified material. This will be helped by a number of rapidly advancing verification techniques.

The 1997 EC labelling guidelines risked damaging EU–US trade relations. The USA has said that it will refuse to yield on the issue of segregation, making the labelling guidelines difficult to implement. In addition, the labelling regulation may be declared illegal under WTO guidelines, which regard 'unreasonable labelling' as a non-tariff barrier to trade, if the USA decides to object to the regulation at any time. However, the EC decision not to extend the ruling to segregation, allowing companies to comply with Directive 90/220/EEC without having to segregate modified produce from unmodified produce, may be enough of a concession to safeguard the mandatory labelling requirement. Meanwhile, there was continuing pressure within Europe for segregation. Agriculture Commissioner Franz Fischler has, however, been unable to persuade US Agriculture Secretary Dan Glickman to accept the principle of segregation, as the USA believes it would pose a threat to its agricultural exports. The EC therefore had to reject segregation in July 1997 due to certain US trade reprisals.

The mandatory labelling rules in the EU were due to come into effect on 1 November 1997, and should have been applied retrospectively to Monsanto's Roundup Ready™ soybeans and Novartis' *B.t.* maize. The EC stalled on implementing the rules, however, under continuing political and commercial pressure. Meanwhile, multinationals had started to make some movement towards labelling in 1997, in anticipation of stricter regulations being enforced. For example, Novartis stated that it would begin labelling some of its genetically modified products. It started by labelling bags of maize seed as to whether or not they contained genetically modified seed. Plant Genetic Systems,[22] as a condition of EU marketing approval in 1997, also started labelling bags of its transgenic canola seed.

Negative labelling and organic food

Even if multinationals, the WTO and free trade agreements ultimately thwart mandatory labelling of all genetically modified foods, recent events have shown that a clear market exists for foods that can be labelled as guaranteed free of genetically modified ingredients or that have been made without the use of genetic engineering. Some sectors of the food industry are worried about negative labelling of this type, which might eat into their market share. Although the major multinationals in the food industry have raised economic arguments against the segregation of commodity crops, other suppliers have been able to obtain guaranteed non-modified supplies of these crops to make products that carry voluntary labels for niche markets.

Consumer groups campaigning for the mandatory labelling of all genetically engineered foods have claimed that these foods are the antithesis of natural foods. Such pressure has helped prevent any food produced from transgenic crops being labelled as 'organic' in Europe[23] and in the USA.[24] Growing transgenic crops under organic conditions, therefore, does not enable them to be labelled as 'organic'. MEPs in May 1997 extended Regulation 2092/91 on the Organic Production of Agricultural Products to cover animal products as well as crops. This meant that any use of genetic engineering in animal or crop production prevented a food being labelled as organic, 'to protect consumers' trust in organic products'.[25] The EU agreed in December 1996 that organic products could be labelled as free of genetically modified organisms (GMO-free).[16] It was proposed that a logo to this effect could appear on organic farm products from January 1988 in the EU. In response to this proposal, the US government and US-based multinationals threatened legal action, through the WTO, if such a GMO-free label is introduced. Previously, Monsanto has attempted to sue companies in the USA who tried to label milk products as rBST-free,[26] when these labels were seen to have a positive effect on rBST-free milk sales.

The July 1997 EC labelling guidelines extended the use of negative labelling, by allowing manufacturers to voluntarily label any food certified as not containing genetically modified ingredients with a 'this does not contain' or 'not containing' label. Meanwhile, retailers around Europe started to extend their voluntary labelling of genetically modified foods. The British Retail Consortium, for example, announced that leading British supermarket chains would from 1998 label products, containing soybean and maize from the USA, as containing genetically modified ingredients.[27]

Mandatory labelling of genetically modified foods has been agreed, at least in Europe, thanks to sustained consumer and political pressure, while retailers are starting to obtain alternative supplies of commodity crops for customers who do not want food made with genetically engineered ingredients. However, the EC's mandatory labelling guidelines are potentially under threat by free trade rulings, and multinationals and the US government continue to resist the crop segregation that would make such labelling most effective, with the result that this gain in consumer choice is still incomplete and could be short-lived.

Notes

1. Webb and Lang, 1990.

2. Piccioni, 1990.

3. A typical supermarket label read: 'Safeway Double Concentrated Tomato Puree. This product is produced from genetically modified tomatoes. Please ask if you require further information.'

4. The *Codex Alimentarius* Commission was established in 1962, by the Food and Agriculture Organization (FAO) of the UN and the World Health Organization (WHO), to set national standards for food, including labelling. Codex had a fairly low profile until the genetically modified food debate.

5. Padgette et al., 1996.

6. *Report of the Committee on the Ethics of Genetic Modification and Food Use, 1993*, HMSO Publications, UK.

7. Fagan, J., 1997, 'Importation of Ciba-Geigy's *B.t.* maize is scientifically indefensible', http://www.netlink.de/gen/BTCorn.htm

8. *AgBiotech: News and Information* 8 (12): 197N, December 1996.

9. Vines, G., 1992, 'Guess what's coming to dinner', *New Scientist*, 14 December 1992.

10. For example: Ladenis, N., letter, *Guardian*, 7 December 1996, p. 22.

11. Goldsmith, 1990.

12. Erlichman, J., 1997, 'A hard price for free trade', *Guardian*, 13 May 1997, p. 17.

13. *Nature* 384: 301, 28 November 1996.

14. *AgBiotech: News and Information* 8 (9): 159N, 1996.

15. *Chemistry & Industry*, 16 December 1996, p. 964.

16. *Nature* 386: 639, 17 April 1997.

17. *Nature* 384: 502-3, 12 December 1996.

18. *ENDS Report* 264, January 1997. From early 1997, Tesco used non-modified soya in all its in-house bakery products; and Tesco and Sainsbury banned unprocessed modified maize as an animal feed ingredient within their meat supply chains.

19. *Europe Environment* 491: 9–11, 14 January 1997.

20. Fischler, F., 1997, 'Genetically modified or not? The answer is the consumers' choice'. http://europa.eu.int/en/comm/dg06/com/htmfiles/welcome.htm

21. *ENDS Report* 270, July 1997.

22. Hoeschst's AgrEvo bought out Plant Genetic Systems in 1996, for US$730 million.

23. *New Scientist*, 28 September 1996, p. 13.

24. *AgBiotech: News and Information* 8 (12): 205N, 1996. However, the USDA announced its intention to redefine 'organic', to incorporate biotechnology, in late 1997.

25. *Europe Environment* 500: 8–9, 27 May 1997.

26. Greenpeace Press Release. 'By Christmas, 60% of your Food could be Genetically Manipulated'. 16 October 1996. http://www.greenpeace.org.uk/science/ge/index.html

27. *New Scientist*, 29 November 1997, p. 5.

14. Impacts on the Third World

The application of genetic engineering to agriculture will have a number of impacts on the Third World. Some of these impacts will be similar to those seen in industrialized countries. Consumers everywhere will be presented with the same potential health risks (see Chapter 8) and the same reluctance to label genetically modified foods (see Chapter 13). Transgenic crops, however, have been developed amidst promises that they will help the Third World feed itself, although this claim seems to ignore the complex social and political factors that contribute to hunger. Meanwhile, markets for economically important Third World agricultural products in industrialized countries are being threatened by alternatives grown using the new biotechnology in tissue culture or in transgenic temperate crops.

Many of the potential ecological risks posed by genetically modified organisms are similar to those in the industrialized nations (see Chapter 7), although in the Third World the clash between traditional agricultural systems and the intensive systems under which transgenic crops grow optimally is more marked. It was claimed that the use of genetic engineering would result in reduced chemical inputs, due to in-built pest and disease resistance. However, the transgenic crops produced to date have not been compatible with sustainable agriculture.

Transgenic crops and the world's hungry

Can genetic engineering be used to help alleviate hunger and starvation in developing countries? The multinational companies involved in producing transgenic crops certainly think so, and use the claim as a selling point for transgenic crops in their promotional literature. However, critics argue that such claims ignore the main causes of hunger and starvation, while pointing out that increasing the amount of food on the planet is not necessarily the solution to feeding the hungry among the world's growing population. Multinationals have concentrated on developing crops that will earn high profits, not crops that could make the best contribution to solving the world's food problems. It is reasonable to expect that companies should strive to maximize their profits, but claims that they are making important contributions to world food supplies are not at present justified.

World food production has been rising by around 1 per cent per annum over recent decades, but the number of people with insufficient food is also growing. Hunger is the result not of insufficient food being grown, but of people being excluded from access to that food. The Food and Agriculture Organization (FAO) of the United Nations estimates that 800 million people have insufficient food to meet their basic needs, while 40 per cent of the world's entire population suffer from malnutrition.[1] Malnutrition is essentially caused by poverty. Poverty is created by a complex mix of social and political factors. Population growth is of great importance, but it is intrinsically interlinked with issues of poverty and food security. The conditions under which widespread poverty in the Third World became established were created during the colonial era. These conditions have been maintained in the post-colonial era by Third World debt,[2] by free trade agreements, by industrial agriculture, which has concentrated on the growing of monocultures of export crops, and by a number of other factors.[3] In some cases transgenic crops are part of the problem of, rather than the solution to, poverty in the Third World.

Africa's ability to feed itself over the past few decades has been steadily diminishing as export crops have displaced food crops grown for local consumption. The production of export crops is not related to the domestic food needs of Third World countries, with the result that crisis-hit countries can have thriving export agriculture sectors. During the Ethiopian famine of 1984, the country was exporting coffee, meat, fruit and vegetables to Europe. While famines occurred in the Sahel countries of Burkina Faso, Mali, Niger, Senegal and Chad in the mid-1980s, these countries were producing record harvests of cotton for export to industrialized nations.[4] The decision to grow cash crops, such as cotton, rather than crops to meet national food requirements is the result of government and aid agency policies. Famines are usually seen as being caused by drought, but agricultural policies are often the root cause of starvation.

In general, the genetic diversity of the world's major crops has been declining, as cash or export crops, usually grown as monocultures, replace local food crops. Monocultures are prone to pest and disease outbreaks because of their genetic uniformity, while genetically diverse crops contain a proportion of plants that are likely to have some degree of resistance to pests and diseases. Monocultures of transgenic plants display a high degree of genetic uniformity, making total crop failure in the face of virulent pests or diseases a possibility. Crop monoculture has been implicated in many cases of crop failure in the past. The Irish potato famine in the 1840s, caused by the potato blight *Phytophthora infestans*, devastated a crop consisting of only one variety and was responsible for the deaths of over a million people. New strains of *P. infestans* threaten crops today. In 1992 the most virulent strain known to date was reported in Mexico, while some strains are now completely resistant to fungicides.[5] Excessive crop uniformity also threatened maize in the USA in 1970. The crop proved highly susceptible to the southern leaf blight fungus (*Helminthosporium maysis*), which spread northwards at a rate of 150 km per day and wiped out 15 per cent of the total US maize crop.[6]

Many countries in the Third World possess the genetic resources to guarantee a sustainable food supply. In India, for example, farmers grow over fifty thousand varieties of rice. In a single village in north-east India, a survey showed that over seventy varieties were being grown.[7] This strategy means that although some varieties may be susceptible to problems of pests or disease in any one year, other varieties will survive. Local varieties, sometimes called landraces, have been bred over many generations in an area and have become adapted to local conditions. Transgenic crops do not have this local adaptation. Traditional plant-breeding solutions in most cases therefore are solving problems adequately, while maintaining sustainability. The suggestion, made by some multinational companies, that a monoculture could be engineered using foreign genes to produce a balanced diet in one particular crop plant that is naturally deficient in certain amino acids (see Chapter 6) is no substitute for a balanced diet based on a diversity of crops. The introduction of such transgenic crop monocultures, with their high genetic uniformity and high chemical input requirements, is more likely to undermine food security and biodiversity.

It is noteworthy that most of the food products from transgenic crops have been marketed to consumers in affluent industrialized countries – for example, Calgene's Flavr Savr™ tomatoes, engineered to cut the cost of manufacturing tomato paste, and Monsanto's high-starch 'quick fry' potato for the fast-food market. Recent developments include a variety of transgenic navy bean (Prim) that causes less flatulence in the consumer, while a range of transgenic fruits engineered for taste may soon reach the market. Tomatoes and other crops that have been genetically modified are already abundant in these Western markets.[8] These are not products designed to alleviate starvation in the Third World.

Multinationals are seeking to grow large areas of transgenic crops in the Third World – for example, tomatoes and potatoes for the major fast-food chains. These crops rarely fit in with traditional local diets or the plans of local farmers. To date, only a handful of crops have been worked on to any great extent: cotton, tobacco, maize, potato, soybean, tomato and canola. Companies have shown little interest in foods grown primarily in the Third World, such as millet, cassava or yams. These crops have constantly been undervalued. Multinationals have also done little work with low-value crops such as wheat or a range of more minor, but still important, vegetable and fruit crops. The production of herbicide-resistant crops has been top of the multinationals research and development agenda, while research on improvements in photosynthesis, nitrogen fixation and drought resistance, which could probably have the most impact on world food production, is still at an early stage.

Research on tropical crops, however, was given a higher profile in 1996 when cassava (*Manihot esculenta*), an important source of dietary carbohydrate in Africa, became the first tropical crop to be given priority for genetic manipulation. Specific techniques for producing transgenic cassava plants were developed at this time.[9] Cassava is difficult to breed by conventional methods and suffers high yield losses due to weeds, pests and pathogens. Researchers aim to produce transgenic varieties with improved starch quality and quantity, increased protein content and decreased levels of cyanogenic glycoside, a toxin that has to be processed out of the roots before they are consumed. Also in 1996, Monsanto started collaborating with scientists in Kenya to produce transgenic sweet potatoes resistant to the feathery mottle virus (FMV).[10]

Most of the major research institutes in developing countries now have biotechnology research programmes. Some of these institutes have extensive collections of major crop varieties – for example, rice at the International Rice Research Institute (IRRI) in the Philippines, potatoes at the International Potato Centre (CIP) in Peru and maize at the International Centre for the Improvement of Maize and Wheat (CIMMYT) in Mexico. These collections represent a valuable genetic resource in developing transgenic varieties. However, these and the other International Agricultural Research Centres around the world have, often in collaboration with multinationals, initially concentrated on developing crops along the lines of the high-yielding varieties of the green revolution, which require high inputs.[11] To make further contributions to meeting the food requirements of local populations, these institutes may increasingly need to adapt biotechnology to meet the particular requirements of their countries.

The initial choice of crops on which to do genetic manipulations was influenced by the fact that researchers use 'model' plants, studied experimentally for many years. The first transgenic plant developed was tobacco, followed by its relatives tomato and potato. These were followed by commodity crops, such as soybeans and maize. Resistance to herbicides has been the predominant modified characteristic, as it represents the most profitable initial genetic engineering application as far as multinational companies are concerned. If the first wave of transgenic crops are a success, however, further beneficial modifications, which are not considered financially viable at the present time, may be commercially produced. Calgene, for instance, has claimed it has had to initially concentrate on commercially viable products for the developed world market just to stay in business. It may, therefore, be harsh to judge the technology by what has come first out of the pipeline.[12] A range of transgenic crops are now in the development stages, which could make valuable contributions to subsistence Third World agriculture. These include drought-resistant varieties and crops resistant to nematodes, insects and other pests and diseases prevalent in the developing world. Techniques developed for increasing the productivity of farmed fish in industrialized countries might also be adapted for species important to the Third World, such as carp (*Cyprinus* spp.), catfish (*Clarius* spp.) and tilapia (*Oreochromis* spp.).

If the application of genetic engineering brings benefits to developing countries, however, it may be difficult to deliver the technology without some loss of autonomy by people in the Third World.[13] The transgenic crops delivered to date have been patented in the industrialized world, will require specific agrochemical inputs that need to be bought from multinational companies, and will be grown at the expense of local crop varieties. To be of great benefit, the emphasis must switch from producing varieties for high-input industrial agriculture, to varieties developed in the Third World and suited to subsistence growing conditions. Ultimately, poverty and hunger cannot be solved by the application of new technologies if the structural, cultural and political aspects of the problem remain unresolved.

Transgenic crops: chemical dependency or sustainable agriculture?

Much of the initial research on genetically modified crops was aimed at making crops resistant to attack by pests and disease, which would reduce pesticide use. This was envisaged as helping make crops compatible with Integrated Pest Management (IPM) practices and biological control programmes. The technology was sold as being a component of a 'green' or sustainable agriculture.

Sustainability is a concept that can be summarized by saying that each generation should pass on to the next a set of undiminished environmental assets, by meeting the needs of the present without compromising the future. Sustainable development is development that maintains an appropriate balance with nature. In sustainable agriculture, crop genetic diversity is preserved, soils remain fertile, pollution is minimized and pest problems are not exacerbated. Sustainable agriculture embraces cultural pest and weed control practices, crop rotation, crop diversification, intercropping, biological control, water conservation, and the recycling of resources and natural fertilizers. Modern organic farming has shown that many of these traditional agricultural ideas can be profitable and compatible with technological advances.

However, genetic engineering is not currently perceived as being compatible with the ideas of sustainable agriculture, largely because of the emphasis on producing herbicide-resistant transgenic crops. It was assumed by many commentators in the early days of genetic engineering that large multinational agrochemical companies would show a guarded response to the use of genetic engineering, which promised reductions in chemical use in crop protection.[14] However, the response of these companies has been to move forward enthusiastically with the development of herbicide-resistant crops. These crops positively demand herbicides for their effective use and a complementary demand can be created for both seeds and agrochemicals, often both supplied by the same multinational company. About 45 per cent of the releases of transgenic crops in European countries between 1992 and 1995 were of plants modified for herbicide resistance.[15] This pattern was repeated worldwide, although the proportion is now decreasing as a number of different characteristics are being engineered into crop plants,[16] although these are often included along with herbicide resistance characters. Herbicide-resistant crops are widely seen as likely to increase herbicide use (see Chapter 4). This will create additional environmental problems as increased spray-drift affects nearby natural habitats and enters streams and rivers.

The use of genes for *Bacillus thuringiensis* (*B.t.*) toxins to create insect-resistant transgenic crops may threaten the usefulness of *B.t.* sprays in IPM programmes (see Chapter 5). *B.t.* sprays are used in low-input sustainable agriculture because they are highly specific to insect pests, thus leaving biological control agents unharmed. If large areas of transgenic crops containing *B.t.* genes were to be grown in an area, however, increased resistance to the toxin would build up in pest insect populations.

The transgenic crops produced to date have been designed for use in high-input industrial farming, like the high-yielding crop varieties of the Green Revolution before them. The Green Revolution was immensely successful in increasing crop yields because of the development of high-yielding crop varieties and the use of chemical inputs, but this resulted in the disruption of many sustainable agricultural practices. Farmers using transgenic varieties risk being caught on a similar chemical treadmill, with crops requiring high chemical inputs to achieve their promised yields, particularly high fertilizer applications. The USA alone currently uses over 12 million tonnes of fertilizer annually. Fertilizer use is increasing dramatically: between 1980 and 1990 fertilizer applications worldwide more than equalled all fertilizer applications made previously in human history. All this fertilizer use is causing environmental problems. Excess nitrogen is responsible for algal blooms, urban smogs, the death of trees, the leaching of nutrients from the soil and the loss of fragile habitats.[17] In common with the high-yielding crop varieties of the Green Revolution, transgenic crops also require high levels of irrigation, a drain on valuable water resources, particularly in developing countries. In addition, they are likely to be intrinsically less resistant to drought, floods and disease than many traditionally grown varieties.

Fertilizer use could potentially be reduced by the development of nitrogen-fixing transgenic crop varieties. The rice varieties that came to predominate during the Green Revolution were selected for their optimum response to artificial fertilizer, with the result that their natural nitrogen-fixing capability, via *Rhizobium* bacteria on their roots, was selected against. Transgenic crops need not necessarily carry on creating the negative effects of the Green Revolution, but could be used to solve some of them. For example, drought-tolerant varieties would help to conserve water by requiring less irrigation.

Transgenic crops are, however, in many cases unlikely to represent the most appropriate technology for food production. Traditional growing practices involving intercropping or multiple cropping and crop rotation are effective at keeping in check a range of pest, disease and weed problems. Multiple cropping is successfully practised in many developing countries, where the techniques originated many years before pesticides started to be used. In Africa, 98 per cent of cowpeas are grown in combination with other crops, while in Nigeria, over 80 per cent of all cropland is given over to multiple cropping.[11] The advantages are great, especially for small farmers, and it represents a highly effective and cost-free form of pest control. Populations of insects are frequently lower on crops grown in mixed cropping situations compared to monocultures, and problems due to fungal and viral diseases are also reduced. There is often a substantial increase in combined crop yields compared to monocultures, a more optimum use is made of available land and environmental resources, weed invasion is reduced due to greater soil coverage, and there is less soil erosion. Increased soil fertility can also result and external chemical inputs are reduced.[11] IPM, incorporating biological control and reduced pesticide inputs, represents an alternative approach that has made big advances in controlling pests over recent years in developing countries. IPM can be made compatible with traditional growing practices.

Critics of genetic engineering applications in agriculture view them as hi-tech, high-input strategies that represent just another 'technological fix'. Single gene transfer solutions are offered to a range of problems, for which traditional agriculture often already has answers. The danger for developing countries is that transgenic crops might come to replace the more appropriate technology of traditional methods. Transgenic crops are strongly promoted as being scientifically advanced and superior to previous varieties, and traditional practices may as a result come to be erroneously viewed as backward in some way. In reality, intercropping, crop rotations, genetic diversity, indigenous resistant varieties and native insecticidal plants can provide adequate pest and disease control in most cases.

If transgenic crops are developed to thrive on degraded soils, and in drought conditions they may provide great benefits, there is a danger that they might cause complacency and lead to the sources of environmental degradation being ignored. In this scenario, farmers will wait for the next technological fix to solve the situation, while the environment degrades further. The application of technological fixes is no substitute for confronting the underlying reasons for ecological damage.

Transgenic crops may come to have an adverse effect on biodiversity, as previously mentioned. The technology represents a solution that aims to produce a small number of super-varieties engineered to solve particular problems. Transgenic crops, with a high genetic uniformity, may be promoted in preference to traditional varieties because of the presence of proprietary genes making them more profitable. There are also concerns about the declining diversity of livestock, a trend that might be exacerbated by transgenic animal production. A FAO report concluded that domestic livestock breeds are disappearing worldwide at an annual rate of 5 per cent, or six breeds a month.[18] As with transgenic crops, genetically modified animals are designed for a high-input, intensive and industrialized agriculture.

The perception that transgenic plants will lead to increased agrochemical dependency is ironic, as their development initially offered genuine potential for a crop protection that was more environmentally friendly, with reductions in chemical inputs. Reduced pesticide use could have led to reductions in health risks to farm workers, less chemical run-off to rivers and fewer pesticide residues in foods. However, the promise that transgenic crops would be a useful component in IPM programmes and of value to sustainable agricultural systems has not been fulfilled.

Economic impacts

The transfer of genetic engineering technology to the Third World will in most cases involve multinational companies, many of whom have sales figures that are comparable to the gross national product (GNP) of the developing countries in which they operate. A number of the key disputed issues relating to the activities of multinationals in the Third World are relevant to the discussion of genetic engineering in agriculture.[19] These issues concern whether capital is being brought into the country in whatever form; whether the technology is contributing to a widening of the gap between rich and poor; the extent to which the technology is being transferred (e.g. how much research and development is being done in the developing country and whether the company retains monopoly control over the technology); whether indigenous products are being displaced; and whether the technology is appropriate, or to what extent it is adapted to local factors.

Farmers in developing countries will hope to increase their incomes by growing transgenic crops. However, they will have to pay more for transgenic seed. Although large-scale farming operations in developing countries stand to benefit from the technology, most farmers in the Third World may not because they cannot afford the extra costs, both of the transgenic seeds and of the other inputs required, including fertilizer, irrigation and herbicides. Although available to large landowners, these inputs are beyond the means of many small-scale farmers. Therefore, the technology as it currently stands is not decreasing the gap between the rich and poor, and is not being adapted to suit the local conditions found in developing countries. If this trend continues, the benefits of this 'gene revolution' in agriculture will not be felt by those who really need additional food. Governments need to intervene in some way if the benefits are to be spread to small-scale and subsistence farmers.

Multinational companies fund most of the research in biotechnology and transgenic crops, and increasingly direct the types of research programmes conducted in universities. The flow of technical information to the Third World is slowing because of concerns regarding the patenting of transgenic crops. The development of transgenic organisms has been done almost exclusively in the laboratories of multinationals in industrialized countries, while patent protection means that these companies keep tight control over their transgenic seeds. If drought- or pest-resistant varieties were developed by multinationals for developing world markets, they would be patented and could only be grown under licence. Traditional growing practices often involve farmers keeping back seeds from one year's crop, in order to plant them the following year. However, additional royalty payments are required to be paid to multinational companies if further generations of seed are grown from patented transgenic crops. Gene-licensing agreements may in fact prohibit farmers keeping back seed from certain crops. While multinationals actively promote their hi-tech transgenic seeds, traditional crop varieties are becoming undervalued. Yet local varieties are often adapted to local growing conditions, as previously noted, and are not subject to patent protection.

Farmers in the Third World may also find their markets shrinking in the face of competition from alternatives produced in industrialized countries, either grown in temperate transgenic crops or produced using genetically modified microbes. Calgene, for example, has produced a genetically modified canola, or spring rape, containing high levels of lauric fatty acids, used in the manufacture of soaps, shampoos, detergents and confectionery. These oils are traditionally derived from coconut and palm kernel oils, and not normally obtained from any non-tropical plant. The Philippines is the world's largest exporter of coconut oil, which accounts for 7 per cent of the country's total export earnings. The crop provides direct or indirect employment for 21 million Filipinos, about 30 per cent of the country's population. A RAFI report[20] suggested that the economies of the Philippines, and other coconut oil-exporting countries such as Indonesia and Malaysia, could be severely affected by any large-scale plantings of transgenic canola in North America or Europe. The USA is the largest importer of lauric acid. In 1995, over 8,000 hectares of high-lauric canola were grown in southeastern America. In 1997, this had risen to 28,000 hectares, with further increases likely. Proctor and Gamble, one of the world's largest buyers of lauric acid, has agreed to buy lauric acid from transgenic canola. Calgene's high-lauric canola is only the first in a line of value-added proprietary plant oils; by 1996 Calgene held 45 patents concerning plant oils.

In the short term, however, exports of palm oil from Malaysia to Europe have increased, due to demand for alternatives to unsegregated soybean oils containing genetically modified soybeans. Continued consumer opposition in Europe to foods produced from transgenic crops is therefore aiding Third World countries as they adjust to the rapid changes in agricultural production initiated by biotechnology.

Other tropical crops whose markets are vulnerable due to developments in biotechnology include vanilla, chocolate and sugar. Production of these is shifting from the tropics to laboratories in industrialized countries. Vanilla (*Vanilla planifolia*) is a major export crop for Madagascar, the Comoro Islands and Réunion, which between them account for around 98 per cent of the world's vanilla production.[21] In Madagascar, over seventy thousand smallholders are involved in the growing of vanilla and the crop accounts for 10 per cent of the country's export earnings; in the Comoro Islands, vanilla accounts for 60 per cent of foreign exchange earnings.[21] However, vanilla can now be produced by taking plant tissue and growing it under tissue culture conditions, which threatens the economies of these countries. The USA is the world's largest importer of vanilla. American companies, including David Michael and the Escagenetics Corporation, have developed the techniques, which are undercutting the agricultural producers.[21] David Michael, a company based in Philadelphia which collaborates with the University of Delaware, is also breeding hardier strains of vanilla to extend its growing range. Escagenetics, based in California, claims that it can produce vanilla at a fraction of the cost of natural vanilla extract. The company has applied for patents on its 'Phytovanilla' and 'Phytovanillin' products.

Cocoa (*Theobroma cacao*) is the second most important tropical commodity in international trade. Most of the world's cocoa is grown in West Africa and nearly half of it is grown by smallholders. American companies, including DNA Plant Technology and Hershey Foods, are developing new cocoa varieties using tissue culture techniques. Research in the USA and Japan is leading to the production of cocoa butters, for chocolate manufacture, by enzymatically transforming cheaper vegetable oils or growing yeasts with modified fatty acids.[21] These developments are likely to have a major impact on the African economy.

About 60 per cent of the world's sugar comes from sugar cane (*Saccharum officinarium*), which is predominantly produced in the Third World. Sugar exports are an important part of the economies of many developing countries, for example, many Caribbean islands export over 70 per cent of their production and are highly dependent on the sugar crop. However, during the 1980s the price of sugar cane on the world market collapsed, with little hope of recovery.[11] The decline of sugar cane was precipitated by the EC favouring sugar beet and subsidizing to beet growers, turning Europe into a net exporter of sugar. Biotechnology has now hastened this decline, with the production of sugar substitutes and new sweeteners. Starch-derived sweeteners – for example, high fructose corn syrup, made from maize – were the first alternatives to sugar produced using enzyme technology. Archer Daniels Midland is one of the major producers of corn sweeteners, which are promoted in the USA at the expense of foreign sugar imports. Corn sweeteners are now used in over 95 per cent of the soft drinks sold in the USA.[11] A range of chemically synthesized alternatives to sugar subsequently reached the market, including aspartame, which was patented by G.D. Searle in 1974 and sold under the brand-name NutraSweet. The Nutra-Sweet Kelco Company, a unit of Monsanto, now manufactures and markets NutraSweet brand sweeteners and related food ingredients.

The most serious threat to tropical sugar cane crops now, however, is the development of proteins with many times the sweetness of sugar. These proteins have a strong binding affinity with the sweet receptors on the human tongue, which accounts for their intense sweetness. Thaumatin is a protein obtained from *Thaumatacoccus danielli*, known locally as katemfe in the regions of Western and Central Africa, where it is native. It is 2,500 to 3,000 times sweeter than sucrose. Thaumatin was first marketed by Tate and Lyle under the trade name Talin.[22] Unilever was the first company to isolate the gene coding for thaumatin and insert it into a bacteria. The gene has now been engineered into bacteria and yeast, for the production of thaumatin in fermenters, by a number of companies. For example, Ingene produced a recombinant thaumatin product using the yeast *Saccharomyces cerevisiae.* The thaumatin gene has also been expressed in tobacco and food crops, by a number of different multinationals seeking ways of obtaining cheap plentiful supplies of the protein. The prospect of making edible crops sweeter is now also a possibility using this technology.

Gene hunters from the multinational companies have been scouring the Third World for similar proteins to thaumatin.[21] For example, there is interest in a plant, *Lippia dulcis*, which the Mexican Indians have chewed for centuries,[11] and a subsidiary of Suzuki in Japan is marketing proteins from bertoni (*Stevia rebaudiana*), which grows in Paraguay and parts of south-east Asia, as a food ingredient. The University of California is using genetically engineered cell cultures to produce the protein monellin, which is 3,000 times sweeter than sucrose, from a gene extracted from the serendipity berry (*Dioscoreophyllum cumminisii*) found in West Africa.[11] The Kirin Brewery Company in Japan is also producing industrial quantities of monellin, by engineering several copies of a modified form of the serendipity berry gene into the yeast *Candida utilis*.[18] Another molecule of interest is the glycoprotein miraculin, isolated from the plant *Richardella dulcifera*. Miraculin is not in itself sweet, but has potential in genetically modified foods because of its effect on the taste receptors in turning sour tastes to sweet – for example, lemon tastes of orange.

A number of other protein sweeteners are now being developed. Some are reported to be around 7,500 times sweeter than sucrose.[23] A drawback with many of them is that they have a bitter aftertaste, but this is being remedied using protein engineering. These molecules are so sweet they can be added to processed foods without increasing the calorific value of the product. The molecules and processes are being patented by multinationals in the industrialized nations. For example, Beatrice Foods in the USA, which funded the research by Ingene on the isolation and cloning of the thaumatin gene from katemfe, owns patents on the gene and its expression in yeast. The use of genetic engineering in the production of protein sweeteners could have devastating effects on the sugar cane production in the Third World, where many millions of people still depend on sugar export markets for their livelihoods.

A range of important export crops grown in the Third World, therefore, will soon be competing with alternatives produced using biotechnology and genetic engineering in industrialized nations. The advantages of land and the appropriate climate are no longer key factors in the production of oils, flavourings, sweeteners and numerous other products from tropical crops. Countries with technological and scientific knowledge will soon corner new agricultural markets, while many developing countries will suffer massive loss of export earnings. Developing countries need to diversify away from single export crops (for example, Madagascar relies on vanilla exports), but now that so many agricultural products can be manufactured using genetic engineering technology, the options for tropical agriculture may become increasingly limited. Over the long term, the new biotechnology may result in a significant relocation of agricultural production out of developing countries, worsening their trading positions, indebtedness and general dependency on the industrialized nations. Even if developing countries can overcome the obstacles, such as patent restrictions, and establish production of sweeteners and flavourings using biotechnology, millions of agricultural jobs would still be at risk.[24]

The products of genetic engineering are often aggressively marketed in the Third World. Monsanto's recombinant BST was marketed to dairy farmers at low prices in Mexico, where its use did not meet the same level of consumer resistance as in many industrial countries. Multinational companies, most notably the GenPharm company, also plan to market in the Third World infant formula milk that has been nutritionally enhanced using genetic engineering. The powdered milk will be produced by transgenic cows that express high levels of lactoferrin protein in their milk. This milk is likely to bring benefit to premature babies, but however well milk from transgenic cows mimics the composition of human milk, it still carries risks from water contamination and will provide none of the important immunological benefits of breast-feeding. The World Health Organzation recommends breast-feeding for two years and beyond in developing countries for these health reasons.[25] Infant formula must be mixed with water, which can easily become contaminated with disease-causing microbes. Breast-fed babies suffer less from diarrhoea, meningitis, and gut, ear, respiratory and urinary infections than bottle-fed babies. Oxfam estimates that around 1.5 million babies a year die from unsafe bottle-feeding. Breast-fed babies also gain protection from antibodies found in human milk, which can attack microbes directly or prevent them from penetrating tissue. Factors in human milk also appear to induce the immune system to mature more quickly.[25] Multinationals have in the past mounted promotional campaigns for infant formula in developing countries that have run counter to the breast-feeding messages of health and aid organizations. Nestlé products were boycotted between 1977 and 1984 in response to their aggressive marketing of infant formula in the Third World. An International Code of Marketing Breast-Milk Substitutes was adopted in 1981 by WHO, which recognized that a legitimate market existed for infant formula, but which restricted how it could be advertised and promoted in developing countries. This code has been accepted by Nestlé and other multinationals. However, Oxfam and other non-governmental organizations have reported repeated violations of this code and the subject of marketing infant formula in the Third World remains controversial.

Multinational companies undertaking research and development involving genetically engineered organisms will conduct experiments in countries hospitable to that work. For example, German-based companies have relocated their pharmaceutical operations to the USA because of strict regulations in Germany. In 1986, officials of the Pan American Health Organization carried a viral vaccine from the USA to Argentina in a diplomatic bag, evading Argentinian import laws, to perform an experiment on cattle without informing the appropriate authorities in either Argentina or the USA.[26] Experiments that are outlawed in industrialized countries may increasingly be conducted in the Third World, where regulations are more relaxed, with associated ecological risks if genetically modified organisms persist in the environment.

Far from helping feed the hungry, the new agricultural technologies may be increasing the economic problems of developing countries, exacerbating poverty and malnutrition. If the technology can be adapted to the specific needs of developing countries some transgenic crops might come to make positive contributions to food production, but only if tied to policies of land reform or other social and political changes that favour the distribution of food to those who need it most. The land area available to agriculture in many areas is shrinking due to the processes of desertification. There is a need for new crop varieties that produce good yields, while tolerating drought and poor soil conditions; that are resistant to pests and diseases, and require few pesticide inputs; and that make efficient use of environmental resources without artificial fertilizers. These crops should not, however, be deployed without also tackling the ecological causes of environmental degradation. Genetic engineering has great potential to increase agricultural production and its real challenge may well be to achieve this goal in the developing world, but it only has the potential to solve a small part of the problem, and this should be borne in mind when evaluating the potential benefits and risks of releasing transgenic organisms to the environment.

Notes

1. Report of FAO/UNFPA Expert Group meeting on food production and population growth, Rome, 3–5 July 1996. http://www.undp.org/popin/fao/expmtg.htm
2. Third World debt in 1997 was estimated at over US$2,000 billion, or twice as much as in 1990. Third World countries pay more in servicing debts than they receive in aid. As most developing countries cannot pay the interest, the crisis grows. *Observer*, 20 July 1997, p. 5.
3. Raghavan, 1990.
4. Timberlake, 1985.
5. Edmunds, R., 1996, 'Tomorrow's bitter harvest', *New Scientist*, 17 August 1996, pp. 14–15.
6. Tudge, 1988.
7. Ho, 1996.
8. Mellon, M., 1996, 'Ripen-on-command: in a society with ample food, why bother?', *Nature Biotechnology* 14: 800, July 1996.
9. Schöpke et al., 1996.
10. *NewBioNews: Information on Biotechnology*, 1996, Monsanto.
11. Hobbelink, 1991.

12. Schmidt, K., 1995, 'Whatever happened to the gene revolution?', *New Scientist*, 7 January 1995, pp. 21–5.
13. Tudge, 1993.
14. Meeusen and Warren, 1989.
15. Landsmann and Shah, 1995.
16. Goy and Duesing, 1995.
17. Pearce, F., 1997, 'Planet earth is drowning in nitrogen', *New Scientist*, 12 April 1997, p. 10.
18. RAFI, 1997.
19. Todaro, 1989.
20. RAFI, 1995. Calgene's canola produces 45 per cent laurate by weight in seed oil. It was created by transferring a gene, expressing the enzyme thioesterase, from California Bay Laurel. Calgene markets high-lauric oil under the brand-name 'Laurical'.
21. Juma, 1989.
22. Thaumatin purified from natural sources was marketed by Tate and Lyle under the brand-name Talin in the 1980s. It is now marketed by Hays Ingredients under the same brand-name. It is widely used in soft drinks, chewing gum and other products.
23. *New Scientist*, 10 May 1997, p. 28.
24. Kennedy, 1993.
25. Newman, J., 1995, 'How breast milk protects newborns', *Scientific American*, December 1995.
26. Wheale and McNally, 1990.

15. Prospects for genetically modified foods

If opinion polls are to be believed, consumers are becoming increasingly suspicious of genetically modified foods. Consumers will ultimately determine how successful these foods are in the marketplace, through their buying decisions and via pressure on food retailers and governments. The outcome of this may also determine to what extent agricultural applications of genetic engineering are developed worldwide. For example, multinationals have hinted that profits need to be made on high-value products for markets in the industrialized nations before investments are made in transgenic crops for food production in the Third World.

Who benefits?

To understand how genetically modified foods so quickly became part of our diet, it is instructive to summarize who benefits from them. This revolution in food production is ultimately driven by economic factors. It has been estimated, for instance, that the potential market for biotechnology-related products within the European Union (EU) will be US$278 billion by the year 2000, with up to 70 per cent of this growth coming from the agriculture and food sector.[1]

Multinational companies benefit in a number of ways from the development and sale of genetically modified food. A modern multinational company involved in this area is typically an amalgamation of a number of the following: an agrochemical company, a seed company, a food-processing company, a veterinary products or a pharmaceutical company. DNA is common to all these areas and developments can be used broadly in many fields. Research breakthroughs in, for example, tissue culture or gene-transfer technology can be exploited in different divisions, such as food production and pharmaceuticals. Developments in one part of the company structure can also be used to generate markets and profits for another part of the company. For example, herbicide-resistant transgenic crops help to sell more of the same company's herbicide. Even if no extra herbicide is sprayed on crops, as multinationals working in this area still claim, gene-licensing agreements ensure that farmers spray only agrochemicals approved by that company on transgenic crops.

The main profit from the development of transgenic crops is the sale of seed, which can be sold at a premium price with subsequent royalty payments. Transgenic crops produced to date represent a continuation of the high-input and high-yield varieties of the Green Revolution. These seeds require high inputs of agrochemicals and fertilizer in order to achieve their high yields. Farmers in the USA spend about US$10.5 billion a year on fertilizers, approximately US$9.4 billion on farm equipment and over US$3 billion on seeds of the major food crops. The US market for agrochemicals reached a record US$10.4 billion in 1995.[2] Transgenic crop varieties are unlikely to reduce these figures.

Farmers should reap major benefits from transgenic crops in the short term, by decreased weed, pest or disease problems leading to substantial profits. Transgenic *B.t.* maize and cotton result in less expenditure on insecticides. Herbicide-resistant crops reduce yield loss due to weeds. The high uptake of transgenic seed by farmers, and their willingness to pay premium prices, illustrates the enthusiasm of many growers for these new varieties. Genetic engineering technology, however, does not at present represent a sustainable solution to agricultural problems (see Chapter 14). Farmers risk being caught in a treadmill, where they become increasingly reliant on chemical inputs. Farmers planting transgenic seed also lose rights over patented transgenic seed and are losing control of how crops are grown, as companies supplying transgenic seed start to dictate the levels of particular brand-name insecticide, herbicide and fungicide inputs, fertilizer applications, the row spacing, amount of irrigation and harvesting techniques. They are in effect being told how to farm.

Food-processing companies benefit from an abundant supply of raw materials that have been designed to suit their needs. Transgenic tomatoes and potatoes have been produced with a higher solid content, so greater profits can be made from the manufacture and sale of tomato purée and French fries. Potatoes are being increasingly grown in tropical highland regions. By the year 2000, it is estimated that a third of the world's potatoes will be grown in developing countries, compared to just 4 per cent in 1950.[3] The priority in the developing world, according to the director-general of the International Potato Centre (CIP) in Peru, is to develop varieties to meet the needs of the fast-food industry, as companies like McDonalds expand into the Third World.[4] Transgenic fruit and vegetables are produced that take longer to ripen and to rot, so less food is spoiled before it can be processed. Genetic engineering is therefore already contributing to the more economical production of processed food.

Supermarkets also benefit from longer shelf-life fruit and vegetable produce, with potential reductions in wastage, although transgenic crops have to date been grown mainly for the food processing industry. However, transgenic fresh fruit and vegetables require labelling as being genetically modified. The labelling of fresh food items, such as fruits, vegetables and fish, which often have a 'healthier' image than processed foods, may be detrimental to sales. It has been suggested that transgenic fish may even have a negative effect on the marketing image of fresh fish as a whole.

Retailers have, in addition, been frustrated by not knowing which processed foods contain genetically modified ingredients. This was particularly acute in 1996 with imports of mixed consignments of soya and maize from the USA. Some retailers have changed their suppliers in cases where they want to ensure that products do not contain food produced using genetic engineering, for example, in their own-label brands.

Farmers, paying more for transgenic seeds, are likely to pass the extra costs on to consumers. This would be in contrast to a general decline in food prices since the Second World War.[5] Food-processors and retailers may pass on some of the savings made from longer shelf-life produce, but in general the introduction of genetic engineering technology is unlikely to make food cheaper. So how do consumers benefit? Consumers are usually unaware that they are eating genetically modified soya, maize or oilseed products in processed food (or were drinking cows' milk produced from rBST-treated herds in the late 1980s) because they are added to general commodity stocks of food and are not labelled. Consumers are not necessarily getting a better quality product. In fact, multinationals insist that modified commodity crops, such as soya, are not different from unmodified supplies of the same crop. Consumers are therefore getting little essentially new, as the foods that have been modified to date are already abundant, and no nutritional or cost-cutting benefits.

On the other hand, consumers will soon have a choice of designer foods, modified for flavour, odour, composition, shape and other attributes. These genetically modified foods will be heavily marketed and their novelty value will ensure healthy profits, in the short term at least. They will represent a relatively small proportion of genetically engineered food, and are more likely to require specific labelling in markets around the world. These designer foods will be the most visible use of the new biotechnology in food production. The Flavr Savr™ tomato will be followed, for example, by a range of other slow-ripening fruit, low-fat quick-fry chips and vegetables and fruits engineered to contain sweetness genes. The question may be whether consumers consider these modifications sufficient reason for meddling with food in such a radical way.

Perceived risks and benefits

The general public's acceptance of genetically modified foods may rest on a perception of risks and benefits. Do the benefits of these foods outweigh the potential risks they pose? A balance of risks and benefits has been used to compare a range of applications of genetic engineering, in surveys designed to see what uses the public would like to see the new biotechnology put to.

Risk is an estimation of the chance that undesirable effects may occur and it is normally statistically assessed from past experience. This 'familiarity principle', however, is not present in assessing the risks of genetic engineering. In the light of recent research, there might not be enough knowledge or sufficient understanding of gene regulation to predict the risks genetically modified crops pose. A case-by-case approach is needed. Even if risk could be estimated from scientific data, it is likely to be at odds with the public's perception of the risks regarding genetically modified food. The perceived risks are likely to be influenced by the fact that mistakes may have important and irreversible consequences for the environment, and therefore even if risks are low they are considered of importance. A number of risks associated with genetically modified foods have been identified, including the spread of transgenes in the environment (see Chapter 7) and the potential development of antibiotic resistance in bacteria living in the human gut (see Chapter 8). The ecological risks, in particular, suggest that very good reasons are needed to justify the genetic modification of organisms.

People can overcome their initial resistance to technology when that technology meets what they consider as important unmet needs. The foods produced to date using genetic engineering have, however, been modifications to foods that are already abundant and of a high quality. Modifications to improve the taste of vegetables may be viewed with scepticism, as traditional vegetable varieties are often perceived as tasting better. Modifications for health reasons, such as altering fatty acid composition, are no real substitute for a change of diet if health is of genuine concern. This contrasts with the use of genetic engineering in medicine, where the technology is seen to fulfil an urgent need. A number of polls have shown a higher level of public support for genetic engineering applications that produce, for example, life-saving drugs, than for food production applications. Within food production, transgenic animals are usually seen as being less acceptable than transgenic plants.[6]

In a poll conducted in Europe in 1996, insect- and disease-resistant crops were perceived as more useful, less risky and more morally acceptable than foods produced with a longer shelf-life or modified for taste or biochemical composition.[7] However, all food-related applications were again scored lower than medical applications.

The use of genetic engineering in livestock production in industrialized countries is regarded by many as unnecessary, as milk and meat are already produced in sufficient quantities, or even over-produced. According to critics, the technology serves primarily as a profit-making exercise for multinational companies. For many consumers it is morally questionable because the ends do not justify the means, which are perceived as causing increased stress to animals. The development of transgenic animals for the production of therapeutic drugs in their milk has clearer perceived benefits, fulfilling definite medical needs. However, these therapeutic drugs could also be produced in bacteria using biotechnological processes.

The potential risks of transgenic foods are in many cases balanced against seemingly small benefits for the consumer, although the benefits to multinationals, growers and food-producers may have knock-on effects in terms of the economy, decreased wastage of food resources and via a range of other factors. It is also worth noting that even if the risks of transgenic crop technology are shown to be negligible, that is not sufficient in itself to ensure that the technology is accepted by the public. Food irradiation was concluded to be safe by a range of advisory committees, but was still rejected by consumers.

Given that the benefit to consumers of most genetically engineered food is small, the accuracy and amount of information available becomes crucial for an assessment of risk perception. However, most of the recent developments in biotechnology and genetic engineering are surrounded in secrecy, to protect commercial interests. Multinational companies have made large investments in genetic engineering technology and they use intellectual property right laws to protect these investments. The patenting of techniques, genes and transgenic organisms requires that no previous publication be made. Companies will also keep information under wraps so as not to give away details of technical advancements to competitors. In a highly competitive and fast-moving market, commercial secrecy is seen as being of paramount importance.

A number of consumer groups are worried about the lack of independent assessment of company data submitted for marketing approval of genetically modified foods. This is compounded by the fact that the data are not usually available for public scrutiny. Independent assessment of company data, when it is possible, has occasionally revealed discrepancies between the data and the official conclusions – for example, Monsanto's mastitis data for rBST-treated cows (see Chapter 3). Irregularities were also revealed in data submitted by the pharmaceutical company G.D. Searle in 1974 for the patenting of the artificial sweetener aspartame. An independent examination of the data found that their conclusions significantly underestimated the possible toxicity of aspartame.[8] The US Freedom of Information Act enabled concerned scientists to gain access to Searle's data. This would not have been possible in most countries, like Britain, that lack an equivalent act.

There is a perception by the public that genetic engineering is to some extent risky. The general climate of secrecy for commercial reasons, the lack of segregation and labelling of modified foods, and poor public relations on the part of the food industry, has done nothing to help dispel concerns about genetically modified foods.

The battle for hearts and minds

Multinational companies involved in the food industry thought that the application of genetic engineering would be greeted as good news, because of what they considered as its beneficial effects for crop production and the environment. However, they discovered to their surprise that the products of genetic engineering were viewed by many as contaminated food. These companies then thought that if they could inform the public about biotechnology and genetic engineering, people would come around to their view. However, these perfectly rational consumers who will, when confronted with the food industry's facts and logic, be won over to the cause of genetically modified foods, may not exist. Whether the multinationals like it or not, this is an emotive issue that taps into deep uncertainties about how far humans should meddle with the processes of life. Opinion polls have identified a mismatch between the industry regulators' concern with safety and risk, and the public's concern for the moral acceptability of genetic engineering applications.[7]

A major problem for the biotechnology industry, according to polls, is that the information it provides is just not believed by a large number of consumers. Surveys also show the industry is not getting its message across effectively, unlike pressure groups operating on a fraction of its budgets. In a widely reported survey conducted by the Office of Technology Assessment (OTA) in the USA in 1987, people were asked how likely they would be to believe statements about the risk of genetically modified organisms made by different groups.[9] University scientists were the group believed most often, followed by public health officials and environmental groups. The groups who were disbelieved the most were the firms making the products and the news media. Similarly, in a more recent survey conducted by Eurobarometer, in the EU in 1996, environmental organizations were the group trusted to tell the truth about transgenic crops, with industry and the media least trusted.[7]

The Eurobarometer survey also looked at the level of biotechnological knowledge of the respondents and found that level of knowledge was poorly correlated with support for biotechnology applications.[7] A number of other recent surveys have shown a similar result, including a UK survey, commissioned by the Department of Trade and Industry, of people living in areas close to trials of transgenic crops. In addition, the polls from 1996 showed no further acceptance of genetic engineering than polls from the late 1980s and early 1990s, when the public knew less about genetic technologies. The results of these polls seem to undermine the industry's belief that people are more likely to accept genetic engineering when they learn more about it.[10]

Consumers often feel helpless in the face of major changes being made to the food supply, particularly when they do not understand the reasons for these changes and have no way of effectively making their concerns heard. In Denmark, public resistance to biotechnology declined after the government passed a law requiring industry and government departments to consult the public over proposed regulations covering biotechnology and genetic engineering: a demonstration of how involvement in the decision-making process can make people less antagonistic to biotechnology.[10] This can be contrasted with the situation in other EU countries, where public concerns are given less regard. For example, in the UK a loophole in the government's consultation process allows companies to proceed with field trials of transgenic crops before the public consultation time limit has elapsed. This occurred when individuals writing with their concerns about an AgrEvo trial, of glufosinate ammonium resistant sugar beet in Suffolk, within a time limit that expired on 21 March 1997, were told that government consent had already been given on 17 March.[11] AgrEvo had acted within the legislation, which, for instance, requires them to advertise trials in the local press, but was able to 'fast-track' its application because its field trial was similar to previously approved trails. This suggests that the public will find it increasingly difficult to object to the growing of transgenic crops as it becomes more common.

Companies now know they have a tough job marketing foods that have been produced using genetic engineering. It is understandable that they concentrate on the positive aspects of the technology, but they have been accused of using outdated scientific terms to promote genetic engineering, giving the public the impression that the technology is better understood, safer and more predictable than it really is. Mae Wan Ho, from the Open University in England, has argued that the use of genetic engineering in agriculture has been promoted using simplistic assumptions about genetics, based around an outdated model in which genes on stable genomes determine characters in linear, unidirectional and additive ways.[12] Research has shown the genome to be more fluid and dynamic than this, through various mechanisms by which DNA naturally re-assorts and through complex interactions of genes acting as if in networks (see Chapter 2).[12] For example, transgenes act like naturally occurring mobile DNA fragments (transposons) that insert themselves at random around the genome, with the result that affected endogenous genes no longer function correctly. The selling of genetically modified food using a simple view of genetic processes therefore misleads consumers about the potential health and ecological risks of the technology.

It is true that the first-generation techniques for transferring genes are more haphazard than could be guessed from the multinational's promotional literature. Transgenes are expressed in only a small number of organisms subjected to these gene transfer techniques, while low and variable levels of transgene expression occur. The unsatisfactory stability of transgenes, due to the random way they integrate themselves into the genome, is the reason why marker genes are necessary to identify successfully transformed material. The techniques for gene transfer deployed also mean that unpredictable effects of transgene expression may occur.

Many unpredictable effects of releasing genetically modified organisms into the environment have already occurred. The common soil bacteria *Klebsiella planticola*, engineered to produce ethanol from crop wastes, unexpectedly inhibited the growth of wheat seedlings through toxic effects on beneficial soil fungi. A similar adverse effect on soil fungi was observed when the bacteria *Pseudomonas putida* was engineered to degrade the herbicide 2,4-D.[13] Furthermore, ecological risks posed by transgenic crops include the possibility of herbicide resistance genes jumping to weed species. Such possibilities appear greater in the light of recent findings on the dynamic nature of the genome, even though the actual risks might still be small.

The major public relations problems for the multinationals in the biotechnology industry are probably the abundance of the foods that are being genetically modified, and the development of herbicide-resistant crops, which are perceived as increasing the levels of agro-chemicals in the environment.[10] Milk produced using genetically modified BST, fruits grown with ice minus bacteria, fruits and vegetables engineered for longer shelf-life, and farmed fish engineered for faster growth rates are all modifications to products that are already abundant in industrialized countries. The applications of genetic engineering to food production are then perceived on the one hand as relatively trivial, not cost-saving and as having little or no benefit to the consumer, and on the other being risky, unpredictable and environmentally polluting. It has been suggested that the early product choices indicate that little thought was given to which products would increase public confidence in biotechnology. Multinationals are perceived as having rushed into developments, with no long-term planning or strategy, while paying too little attention to possible public concerns.

In the battle for hearts and minds, environmental groups and others concerned about genetic engineering appear to be winning the early rounds. However, the food industry is now fighting back to try to reassure the public of the safety of genetically modified food. A massive restoration of confidence is needed, at least in Europe. This may be difficult if multinationals do not display more openness or submit to wider consultation, and do not compromise on segregation and labelling. The attempt to claim the moral high ground, by Monsanto, Ciba-Geigy and others, because they had science and logic on their side, has clearly done the cause of genetically modified foods no favours.

A more reassuring opinion poll for the multinationals was conducted by the Food and Drink Federation, in the UK in June 1995, which found that 68 per cent of people claimed not to know anything about biotechnology.[14] Most consumers in the UK were shown to be neither wildly enthusiastic nor strongly opposed to foods from transgenic crops, according to their conclusions, which suggested that the public are still waiting to be persuaded one way or the other. The Food and Drink Federation, which promotes genetically modified foods, launched its FoodFuture initiative in 1995 to inform the British public further about genetically modified foods. The emphasis of this and other industry-supporting initiatives at that time was on the continuity between centuries-old biotechnology and crop improvements with genetic engineering, the safety of modified foods, and the fact that foods produced using genetic engineering are identical to foods produced using traditional techniques.[15]

The biggest industry public relations exercise to date in Europe to promote genetically modified foods was initiated in June 1997 with the first public event of EuropaBio, an association of the world's leading multinationals, biotechnology companies and food companies involved with genetic engineering, including Monsanto, Novartis, AgrEvo, Rhône-Poulenc, Nestlé and Unilever.[16] EuropaBio initiated a multimillion dollar campaign to change the public's perception of genetically modified foods. It effectively acknowledged that a drastic public relations overhaul was necessary by hiring the crisis management consultants Burson-Marsteller. It advised the food industry to avoid discussing the risks posed by modified foods and to move away from the logical fact-based approach, that had until then proved unsuccessful. It suggested that the industry should focus instead on symbols and stress concepts such as caring, satisfaction and hope.[17] The food industry was further advised that the best way of eliciting a favourable consumer response was to embrace the regulatory system rather than adopt a confrontational stance. Although EuropaBio has talked favourably of consumer choice and has accepted the stricter July 1997 EC labelling guidelines, the major multinationals behind EuropaBio have also sent a joint letter to US President Bill Clinton, in which they urged him to threaten the EU with sanctions under the WTO to get their crops onto the European market without segregation or labelling.[16]

In contrast to the 1995 Food and Drink Federation poll, which concluded that few people were strongly opposed to genetically modified foods in the UK, an independent report published in March 1997 by Unilever, the Green Alliance and the University of Lancaster showed a 'disturbing degree of latent public unease about genetically modified foods'.[18] This report claimed that 86 per cent of the UK population supported the labelling of genetically modified foods, while few saw advantages in taste (10 per cent), economics (19 per cent) or healthiness (9 per cent). The report also concluded that public concerns were not being addressed by the political or regulatory framework. An appreciation of the public's views is essential if there is to be any claim of 'democracy'.

Opinion polls have been central to the arguments for, but mainly against, genetically modified foods. They are of immense importance within the debate but, as just shown, can give contradictory conclusions.

It is time to look at opinion polls more critically and assess their limitations. Assessing consumer attitudes to new concepts is difficult. Bias can easily occur through the way a question is worded, particularly when the public does not have a clear understanding of a subject. Questions are easily 'loaded' in these polls. Minor changes in the wording of questions can result in apparently large swings of opinion, while the information provided can lead to major changes of perception.[6] For example, Monsanto criticized a poll in the UK in 1988 that claimed that 83 per cent of consumers surveyed opposed BST. The question put was: 'The daily pinta should remain as it is and not come from cows injected with BST.' No explanation was given of what BST was, or how it works. Monsanto claimed that the poll was misleading as its data had shown no differences in milk obtained from BST-treated and untreated cows, therefore, the 'daily pinta' remained the same even when it came from cows injected with BST.[19]

Other limitations of polls are that a small minority of the public, if vocal and committed, can change the attitudes of a majority; that inadequacies in the interpretation of data can occur, with selective use of data providing pressure groups with sound-bites to reinforce their particular view; and that polls are a broad-brush technique that often do not provide detailed information on the perception of an issue.[6] For instance, Joyce Tate, in her critique of opinion polls, differentiates respondents who act through self-interest and are concerned about a specific use of biotechnology in a specific locations (NIMBY, i.e. not in my backyard), and respondents who are motivated by ethical or value considerations and are concerned by all biotechnology on a global basis (NIABY, i.e. not in anyone's backyard). Opinion polls will tend to lump these two very different groups together.[6] Internal inconsistencies also often arise in the published opinion polls. For example, a low proportion of respondents often claim knowledge of the technology, yet many more can give their assessment of the risks posed by the technology.

Attitudes expressed in opinion polls might not always correspond to behaviour – for example, an expression of disapproval of genetically modified food may not correspond to how foods are chosen in a supermarket. Other factors, such as brand-name, price or country of origin, may have a larger influence on buying decisions. Nevertheless, the companies producing genetically modified foods have been keen followers of opinion polls, which give them feedback on the effectiveness of their public relations campaigns. It is clear that the public relations departments have generally been doing a poor job. Multinationals are therefore having to start listening more carefully to criticisms levelled against them, and are changing their public relations approach to try and restore public confidence in genetically modified foods. They are working, for example, with Burson-Marsteller, who regard public opinion or 'the socio-pathology of public outrage' as being malleable on any issue. Media campaigns may switch from factual science-based reports to items stressing the potential benefits of the technology in general and emotive terms, to try and 'fight fire with fire'.[17] This will address the mismatch, mentioned earlier, between risks and logic, on the one hand, and moral acceptability, on the other. It is unclear at the time of writing how far this change of emphasis will influence public opinion.

Despite the limitations of opinion polls, however, the message is often very clear. One of the most resounding cases was a referendum in Austria during April 1997, involving 1.2 million people, who agreed to the following: 'No food from genetic laboratories in Austria; no field trials of genetically manipulated crops in Austria; and no patents on life.' The high turnout, a fifth of the population, and the large majority gave a clear signal to the government that the people did not want genetically engineered food.[20] A range of other polls around Europe have also shown a large degree of public opposition to genetically modified food. Public opposition is also growing in other areas of the world. The first shipments of Monsanto's transgenic soybeans arrived in Australia and New Zealand in November and December 1996. This prompted widespread public opposition. For example, a poll by AGB McClair, commissioned by Greenpeace and other environmental groups, found that 60 per cent of New Zealanders were worried about genetically modified foods.[21] This has led the Australia and New Zealand Food Authority (ANZFA) to propose tougher regulations governing these foods.[22] In future, companies wanting to market modified foods will have to apply to ANZFA for approval.

The growing opposition to genetically modified food can be seen as part of a wider concern about modern farming practices and food production methods. In the UK, for instance, the BSE crisis and a number of food contamination scares have made consumers wary about food safety. Genetically modified foods have been caught up in this attitude swing against industrialized agriculture.

Bovine spongiform encephalopathy (BSE) or 'mad cow disease' was first reported in cattle in 1986, in Britain. The cause of BSE was identified as the cheap protein supplement fed to livestock, which contained the rendered carcasses of sheep, including animals infected with scrapie, a variant of spongiform encephalopathy. The epidemic was escalated by rendering infected cows and feeding them back to other cows. From 1986 to 1988, cows infected with BSE were sent for slaughter and entered the human food chain. The first cases of a new variant of Creutzfeldt-Jakob disease (vCJD), the human form of spongiform encephalopathy, linked to eating infected beef, were reported in 1995. By the summer of 1997, over twenty cases of vCJD had been reported in Britain.[23]

In the case of BSE, economic savings in the production of animal feed led to what was in effect cow cannibalism. This completed a cycle of infection that led to an epidemic of BSE in cattle. By 1989, 160,000 cows were officially reported infected, an underestimate given the long incubation period of the disease. The organism was known to have crossed the species barrier from sheep to cows, but the potential risks of crossing the species barrier to humans were played down by the government. An independent study estimated that 440,000 infected cows had entered the human food chain in the UK by 1989. The public was repeatedly misled over the safety of eating beef. In 1989, an official ban on offal was instated, but it was subsequently revealed that abattoirs had not been strictly enforcing the ban. Infected material was probably getting into the human food chain until the end of 1995.[24] In June 1997, it was suspected that BSE cases might be frequent and widespread, although under-reported, throughout Europe.[25]

The number of cases of food poisoning in Britain has increased by nearly 60 per cent since the early 1980s, a trend common to many industrialized countries. Much of this can be attributed to the intensification of farming. High levels of *Salmonella* in eggs and chickens were reported in the late 1980s; outbreaks of other bacterial infections, including *Listeria* and *Campylobacter*, also increased. The standard of hygiene in British abattoirs has been criticized in a series of reports, and has been linked to outbreaks of severe food poisoning caused by pathogenic strains of bacteria, including *E. coli* 0157.[26]

Consumers are, therefore, likely to be more cautious about food safety than previously. Consumer concerns are now being given a higher priority, at least in the UK. The UK government announced, in January 1997, the formation of a new independent Food Standards Agency to oversee food safety, conceding that government departments were no longer trusted by the public on this issue.[27] In the interim, the Labour government, elected in May 1997, started to move food safety issues from MAFF to the Health Department. MAFF has traditionally played the dual role of promoting food production and the food industry, while also safeguarding public health. It was widely seen as placing public health issues second to commercial interests.

Further unease about modern farming and food production practices in industrialized countries arises from concerns about chemical additives, the possible presence of residual antibiotics and growth hormones in meat, and pesticide residues in vegetables. In addition, many consumer and environmental groups are concerned about the increasing political power of multinationals which, linked with the globalization of free trade agreements and their influence within the WTO, gives them greater influence in decision-making processes concerning agriculture.

The disillusionment with industrial agriculture is reflected in a revival of organic farming in Europe, for example in Germany, Austria and Switzerland. In Denmark, where there is concern about pesticide levels in groundwater, 150 pesticide products became illegal in July 1997. The country subsequently considered whether to go entirely organic and ban all pesticides. Although organic farming accounts for only 1 per cent of the total agricultural production in the EU, the land area cultivated using organic methods has increased tenfold since the mid-1980s. Around fifty thousand agricultural concerns in the EU now use organic methods to farm around 1.2 million hectares.[28] Many European supermarket chains are now stocking an increased amount of organic products. These can often be sold at premium prices, as being grown organically gives them added value. A large increase in the number of people eating a vegetarian diet has also occurred in many industrialized countries, due to concerns about the safety of meat in the wake of the BSE crisis, and concerns about animal welfare in intensive agriculture.

Genetically modified foods appeared to arrive suddenly on the market, by stealth, and now a bewildering array of such foods are in the development stages. The time between scientific discovery and technology transfer is getting progressively shorter. This gives the public less time to evaluate the implications of technological innovations. The use of genetic engineering in food production is spreading at a rate faster than that of public understanding or consent. Social policy, meanwhile, has struggled to keep up with the rapid advances made in the production of genetically modified foods. Legislation is needed to reassure the public, but legislators are aware that they have a difficult task in achieving the right balance; laws that are too strict might stifle progress in biotechnology and those that are too relaxed will not gain the confidence of the public. Society must decide if the benefits of genetically modified foods outweigh their risks to the environment or human health – risks that may be relatively small, but are unpredictable and ecologically irreversible. A period of time is needed to evaluate the wider implications of genetic engineering, including the long-term consequences for agriculture, the environment and human health.

Genetically modified foods are here to stay. In 1997, over 4 million hectares of transgenic crops were grown in the USA and it is estimated that 60 per cent of the crop seed sold in the USA will have genetically modified characteristics by the year 2000. If the trends of 1996 and 1997 were to continue, the majority of processed foods will be made with genetically modified ingredients and soon a large proportion of the diet of people living in industrialized countries will be produced using genetic engineering or contain genetically modified organisms. A massive social protest could slow the rapid spread of this technology in food production. However, although this is starting to happen in some countries, nothing short of a social revolution would be needed to stop it.

Genetic engineering does have the potential for benefiting agricultural production in the industrialized nations and in the Third World, particularly if adapted to a local scale in the latter case. Experimental programmes are under way that aim to produce new crop varieties with extended growing ranges, tolerance of drought and poor soil conditions, and resistance to a range of pests and diseases. If the technology in general is to fulfil its full potential, however, it must have public acceptability, it must have future economic viability and it must have a workable legislation framework. This potential may be threatened if a rush for short-term profits, prompted by corporate greed, leads to over-regulation of the technology. This could have adverse effects on the whole area of new biotechnology, including developments in the Third World.

If genetic engineering is to make an important contribution to food production in the future it should be as a result of open debate and widespread cooperation between industry, academia and governments. The potential risks of genetic pollution of the environment, and risks to human and animal health, require that the genetic engineering is closely monitored. Society must take responsibility for regulation and use value judgements to decide how the technology is used. Market forces should not be the sole factor determining how the technology develops. The public, through referendums and opinion polls, has already let its views on certain applications of genetic engineering be known. In some cases, society might consider that the benefits obtained using genetic engineering do not justify the potential risks. Many foods currently on the market, containing genetically modified ingredients, supply no benefit to consumers, present ecological and health risks that are insufficiently understood, and are not welcomed by the majority of the public in many industrialized countries. If it is to be deployed in food production, genetic engineering should be developed democratically and with the aid of governments, to produce a wide range of agricultural improvements that not only generate profits for their producers, but also supply benefits to people around the world.

Notes

1. *Nature Biotechnology*, August 1997.
2. RAFI, 1997.
3. *New Scientist*, 26 April 1997.
4. Hobbelink, 1991.
5. Tudge, 1993.
6. Tate, 1990.
7. *Nature* 387: 845–7, 26 June 1997. The Eurobarometer poll was conducted during October and November 1996. The total sample was 16,246 people or about 1,000 in each EU country.
8. Millstone, 1986.
9. OTA, 1987.
10. Coghlan, A., 1993, 'Gene industry fails to win hearts and minds', *New Scientist*, June 1993.
11. *Farmers Weekly*, June 1997.
12. Ho, 1996.
13. Juma, 1989.
14. Food and Drink Federation, 1995, *Modern Biotechnology – Towards a Greater Understanding*, Food and Drink Federation, London.
15. *Foodsense: Genetic Modification and Food*, 1995, MAFF, HMSO Publications, London.
16. Greenpeace Press Release. Amsterdam, The Netherlands. 26 June 1997. http://www.greenpeace.org.uk/science/ge/index.html
17. *Guardian*, 6 August, p. 9 and *Guardian*, *G2*, 13 August, pp. 2–3. Burson-Marsteller is best known for its work representing Babcock and Wilcox during the Three Mile Island nuclear crisis in the USA in 1979 and for representing Union Carbide after the Bhopal disaster in India in 1984.
18. *Uncertain World: Genetically Modified Organisms, Food and Public Attitudes in Britain*, University of Lancaster, UK.
19. Deakin, 1990.
20. *Nature* 386: 745, 24 April 1997.
21. Greenpeace Press Release. Auckland, New Zealand. 20 April 1997. http://www.greenpeace.org.uk/science/ge/index.html.
22. *New Scientist*, 5 April 1997, p. 5.
23. *Nature* 389: 448–50; *Nature* 389: 497–501, October 1997.
24. Penman, 1996.
25. *Guardian*, 6 June 1997, p. 7.
26. An outbreak of *E. coli* 0157 poisoning in Scotland in 1997 resulted in 20 deaths and around 500 people becoming ill.
27. The incoming Labour government published a consultative green paper, drawn up by Philip James of Aberdeen University's Rowett Institute, outlining the remit of the new UK Food Standards Agency. This remit included genetically modified foods. A draft bill was expected to be laid before parliament in autumn 1997; in the interim, a joint Food Safety and Standards group, comprising MAFF and Department of Health staff, was in operation. However, the government stalled on the setting up of the agency because of pressure from the food industry, which objected to food nutrition issues coming under the new Agency's remit.
28. In 1997, the UK had 0.3 per cent of land in organic production, compared with around 10 per cent in Germany and Austria.

Abbreviations

ACGM	Advisory Committee on Genetic Modification (UK)
ACNFP	Advisory Committee on Novel Foods and Processes (UK)
ANZFA	Australian and New Zealand Food Authority
APHIS	Animals and Plant Health Inspection Service of the USDA
ACRE	Advisory Committee on Releases to the Environment (UK)
BEUC	European Bureau of Consumer Unions
BIO	Biotechnology Industry Organization (USA)
CGIAR	Consultative Group for International Agricultural Research
CIMMYT	International Centre for the Improvement of Maize and Wheat (Mexico)
CIP	International Potato Centre (Peru)
COT	Committee on Toxicity of Chemicals in Food Consumer Products and the Environment (UK)
EC	European Commission
EU	European Union
EPA	Environmental Protection Agency (USA)
EPO	European Patent Office
FAC	Food Advisory Committee (UK)
FAO	Food and Agriculture Organization (United Nations)
FDA	Food and Drug Administration (USA)
GATT	General Agreement on Tariffs and Trade
GAFTA	Grain and Feed Trade Association
HDRA	Henry Doubleday Research Association (UK)
IRRI	International Rice Research Institute (Philippines)
MAFF	Ministry of Agriculture, Fisheries and Food (UK)
MAI	Multilateral Agreement on Investment
NAFTA	North American Free Trade Agreement

NERC	Natural Environment Research Council (UK)
NIH	National Institutes of Health (USA)
OECD	Organization for Economic Cooperation and Development
OTA	Office of Technology Assessment (USA)
RAFI	Rural Advancement Foundation International
PBRs	Plant Breeders' Rights
PPA	Plant Protection Act
PTO	Patent and Trademark Office (USA)
PVPA	Plant Variety Protection Act
USDA	United States Department of Agriculture
WHO	World Health Organization
WIPO	World Intellectual Property Organization (United Nations)
WTO	World Trade Organization

Glossary

Allele Alternative form of a gene. Within a population there may be several alleles of a gene, each with a unique nucleotide sequence. The combination of alleles will determine the characteristics an organism displays.

Amino acid The basic biochemical unit of a protein. Proteins are comprised of chains of 20 amino acids in varying order.

Allergen A foreign substance that acts as an antigen in causing an inappropriate immune response.

Antibiotic Substance obtained from a micro-organism that destroys or inhibits the growth of other micro-organisms, particularly disease-causing bacteria and fungi.

Antibody A protein manufactured by the immune system in response to the presence of an invading foreign body or antigen. Each antibody fits exactly to an antigen and destroys it.

Antigen A substance that causes the immune system to produce antibodies.

Antisense gene A gene that has been genetically modified so that it is a reverse or complementary copy of an endogenous gene. Each adenine becomes a thymine, each cytosine a guanine and so on. A transcribed antisense DNA results in mRNA complementary to the endogenous mRNA.

Bacteria Major class of prokaryotes. Consisting of a single cell, without a nucleus.

Bacteriophage A virus that is parasitic within a bacterium. The viruses insert genes into a bacteria's genome and are used in genetic engineering as cloning vectors.

Baculovirus Group of viruses specific to insects, used as bio-insecticides.

Base The nitrogenous part of a nucleotide. DNA has four bases: adenine, thymine, cytosine and guanine. In RNA, thymine is replaced by uracil. The sequence of bases determines the genetic code.

Base pairing The chemical linking of two complementary bases. In DNA, adenine pairs with thymine and cytosine with guanine. In RNA, adenine pairs with uracil. Base pairing is responsible for holding the two strands of DNA together to form a double helix.

BST Bovine somatotropin. Hormone produced in pituitary gland of cows, which affects growth and development. Cows given additional BST will have increased milk yields.

B.t. *Bacillus thuringiensis*. A bacterial pathogen of insects, used as a commercial bio-insecticide.

Cell The structural and functional unit of most living organisms. The smallest unit of living matter capable of reproduction. A cell contains DNA and the organelles necessary for energy conversion and protein synthesis.

Central dogma The belief originally held by geneticists that the flow of genetic information can occur only from DNA to RNA to protein. It is now known that information can pass from RNA to DNA and this is exploited by genetic engineers.

Character A distinctive inherited feature of an organism. Organisms in a population may display different aspects of a particular character, e.g. roundness and wrinkledness are aspects of the pea shape character.

Chloroplast Organelle in plant cells containing the pigment chlorophyll. The site of photosynthesis: the synthesis of organic compounds from carbon dioxide and water using the energy of sunlight.

Chromosome A DNA-bearing structure, carrying the genes in a linear sequence.

Clone A group of genetically identical cells or organisms, derived from a single organism.

Conjugation The transfer of genetic material between bacteria.

Complementary Nucleic acids are said to be complementary when their sequence of bases are a reverse copy of each other, e.g. each cytosine corresponds to each guanine. Complementary DNA (cDNA) can be manufactured in the laboratory from mRNA.

Cytoplasm Jelly-like material in which cell organelles are suspended.

Dicotyledonae One of two classes of plants. Distinguished by having two seed leaves (cotyledons). Dicot crops include potato, tomato, beans and sugar beet. Generally, broad-leaved plants.

DNA Deoxyribonucleic acid. A double helix consisting of two complementary chains on which the genetic code is located.

Dominant A dominant gene allele is one that is expressed in preference to other (recessive) alleles of that gene.

Endogenous Naturally occurring within an organism.

Enzyme A protein that acts to regulate or catalyze a biochemical reaction, without itself being altered in the process.

Eukaryotes One of the two major groups of organisms on earth, including all animals, plants, protozoa and fungi. These organisms are distinguished by having a cell nucleus and other membrane-bound organelles. The other group are the prokaryotes.

Exon The region of a gene that is expressed, i.e. its coding sequence.

Expression The manifestation of the genetic code. A gene is said to be expressed when a protein is translated, via RNA, from its coding sequence.

Gene A unit of heredity composed of DNA and located at a specific position on a chromosome. A gene is a sequence of bases, its genetic code, that codes for a particular protein.

Gene pool All the genes and their different alleles that are present in a population of a plant or animal species.

Gene probe A single-stranded DNA or RNA fragment used in genetic engineering to search for a particular gene. The probe has a base sequence complementary to the target sequence and will therefore attach to it by base pairing. Probes are labelled with markers to be easily visible.

Genome The total genetic material of an individual or organism.

Genotype The genetic composition of an organism, i.e. its total combination of gene alleles.

Genetic code The sequence of bases on DNA.

Genetic drift Changes in gene alleles over time due to chance rather than selection.

Herbicide A chemical compound used to kill weeds.

Hormone In animals, a chemical signal secreted into the bloodstream to regulate the function of tissues or organs. In plants, a growth-promoting substance.

Hybrid The offspring of parents that differ in at least one characteristic.

Inbreeding Mating between closely related individuals, the extreme case being self-fertilization. Inbreeding can lead to an increased incidence of harmful characteristics.

Inheritance The transfer of genetic instructions for different characteristics to successive generations, via reproduction.

Insecticide A chemical compound used to kill insects.

Intron The region of a gene that is not expressed as a protein, that is not part of the coding sequence. A region spliced out at the RNA stage before protein synthesis.

Larva The juvenile stage in the life-cycle of insects, other invertebrates, amphibians and fish. The larvae of moths and butterflies are also called caterpillars.

Ligase enzyme An enzyme that catalyses the joining of DNA fragments. Used in genetic engineering, with restriction enzymes, to insert foreign genes into vectors.

Linkage The tendency for two genes situated close to each other on the same chromosome to remain together during reproduction. Foreign genes are linked in transgenic organisms because of their close proximity.

Mitochondria Organelle in plant and animal cells. The site of energy production. Contain their own DNA.

mRNA Messenger RNA is responsible for carrying the genetic code transcribed from DNA to the site of protein synthesis.

Molecule A fundamental chemical unit. Formed by a tight grouping of atoms.

Monocotyledonae One of two classes of plants. Distinguished by having one seed leaf. Monocot crops include maize, wheat, rice and other cereals. Generally, narrow-leaved plants.

Mutation An inheritable change in the genetic material.

Nematode Roundworms in the invertebrate phylum Nematoda. Characteristically have smooth cylindrical bodies that taper at both ends. Nematodes are abundant worldwide and species are free-living or parasitic.

Nucleic acid Complex organic molecule found in living cells, consisting of a chain of nucleotides. There are two types: DNA and RNA.

Nucleotide The basic building block of a nucleic acid molecule. Consists of a phosphate, a sugar (ribose or deoxyribose) and one nitrogenous base.

Nucleus The organelle in plant and animal cells containing DNA.

Organelle A structure within plant or animal cells, e.g. nucleus, chloroplast and mitochondria.

Outbreeding Mating between unrelated or distantly related individuals of a species. Outbreeding populations usually show more variation than inbreeding ones and have a greater potential for adapting to environmental change.

Pathogen An organism that causes disease.

Phenotype The observable characteristics of an organism. Determined by the interaction between its genetic make-up and the environment.

Plasmid A small DNA structure that can exist and replicate independently of the chromosomes. Bacterial plasmids can be transferred between cells and are used as vectors to produce recombinant DNA.

Prokaryote An organism in which the genetic material is not enclosed in a cell nucleus, e.g. bacteria.

Protein A molecule consisting of one or several long chains of amino acids linked in a characteristic sequence.

Protoplast Bacteria or plant cells that have had their cell wall removed.

Protease Any enzyme that acts to break down proteins into amino acids. For example, trypsin. Several proteases acting sequentially are normally required for complete digestion of a protein to its constituent amino acids.

Recessive A recessive allele is not expressed when two different alleles of that gene are present. The character controlled by the recessive allele only appears in the phenotype when two such alleles are present.

Recombinant DNA The hybrid DNA produced by joining pieces of DNA from different organisms.

Regeneration The process of growing a whole plant from a single cell or group of cells in tissue culture.

Restriction enzyme An enzyme that cuts DNA at a specific base sequence.

Retrovirus A virus that converts its RNA into DNA, using the enzyme reverse transcriptase, to enable its genes to become integrated into its host's DNA.

Reverse transcriptase An enzyme, found in retroviruses, that is used in genetic engineering to produce DNA from mRNA.

Ribosome The site of protein synthesis within a cell.

RNA Ribonucleic acid. Concerned with carrying out the genetic instructions on DNA. RNA types include mRNA and tRNA.

Selectable marker A gene in a transgenic organism that allows it to be selected from non-transformed organisms, e.g. an antibiotic resistance gene.

Substrate The substance upon which an enzyme acts.

Transcription The synthesis of mRNA from DNA.

tRNA Transfer RNA is involved in the assembly of amino acids in a protein chain. Each tRNA is specific for an amino acid and has a base sequence complementary to that of the mRNA which it translates.

Transgene A gene from one species that is being expressed within the genome of another species.

Transgenic organism An organism containing a foreign gene or transgene, i.e. a gene from another species.

Translation The conversion of the genetic code on mRNA to a protein.

Transposon A mobile genetic element, which can insert itself randomly at any point around chromosomes or plasmids. Their presence can modify the action of neighbouring genes.

Vector A vehicle for cloning foreign DNA and transferring it between organisms.

Virus A particle consisting of a core of DNA or RNA surrounded by a protein coat. They cannot reproduce on their own, but infect a cell and take over its molecular machinery.

Bibliography

Abel, P. P., R. S. De B. Nelson, N. Hoffmann, S. G. Rogers, R. T. Fraley and R. N. Beachy (1986) 'Delay of disease development in transgenic plants that express the tobacco mosaic virus coat protein gene', *Science* 232: 738–43.

Ager, B. P. (1988) 'The oversight of planned release in the UK', in J. Fiksel and V. T. Covello (eds), *Safety Assurance for Environmental Introductions of Genetically-Engineered Organisms*, NATO ASI Series, vol. G18, Springer-Verlag, Berlin.

Ager, B. P. (1990) 'The regulation of genetic manipulation', in P. Wheale and R. McNally (eds), *The Bio-Revolution*, Pluto Press, London, pp. 164-73.

Aldridge, S. (1994) 'Ethically sensitive genes and the consumer', *Trends in Biotechnology* 12: 71–2.

Aldridge, S. (1996) *The Thread of Life: The Story of Genes and Genetic Engineering*, Cambridge University Press, Cambridge.

Altenbach, S. B., C C. Kuo, K. Staraci, W. Pearson, C. Wainwright, A. Georgescu and J. Townsend (1992) 'Accumulation of a brazil nut albumen in seeds of transgenic canola results in enhanced levels of seed protein methionine', *Plant Molecular Biology* 18: 235–45.

Ayub, R., M. Guis, M. Ben Amor, L. Gillot, J.-P. Roustan, A. Latché, M. Bouzayen and J.-C. Pech (1996) 'Expression of ACC oxidase antisense gene inhibits ripening of canteloupe melon fruits', *Nature Biotechnology* 14: 826–66.

Barry, G. F., D. M. Stark, Y. M. Muskopf, R. E. McKinnie, K. P. Timmermann and G. M. Kishore (1992) 'Improved potato quality through genetic engineering', *American Society of Agronomy, Abstract* 187.

Barton, K. A. and M. J. Miller (1993) 'Production of *Bacillus thuringiensis* insecticidal proteins in plants', in S. D. Kung and R. Wu (eds), *Transgenic Plants Vol 1: Engineering and Utilization*, Academic Press, New York, pp. 297-315.

Barton, K. A., H. R. Whiteley and N. Yang (1987) '*Bacillus thuringiensis* delta endotoxin expressed in transgenic *Nicotiana tabacum* provides resistance to Lepidopteran insects', *Plant Physiology* 85: 1103–9.

Bishop, D. H. L., P. F. Entwistle, I. R. Cameron, C. J. Allen and R. D. Possee (1988) 'Field trials of genetically-engineered baculovirus insecticides', in M. Sussman, C. H. Collins, F. A. Skinner and D. E. Stewart-Tull, *Release of Genetically-engineered Micro-organisms*, Academic Press, London, pp 143–79.

Boulter, D., G. A. Edwards, A. M. R. Gatehouse, J. A. Gatehouse and V. A. Hilder (1990) 'Additive protective effects of different plant-derived insect resistance genes in transgenic tobacco plants', *Crop Protection* 9: 351–4.

Brisson, N., J. Paszkowski, J. R. Penswick, B. Gronenborn, I. Potrykus and T. Hohn (1984) 'Expression of a bacterial gene in plants by using a viral vector', *Nature* 310: 511–14.

Brunner, E. (1990) 'Science, secrecy and BST', in P. Wheale and R. McNally (eds), *The Bio-Revolution*, Pluto Press, London, pp. 74-81.

Bryant, J. and S. Leather (1992) 'Removal of selectable marker genes from transgenic plants: needless sophistication or social necessity?', *Trends in Biotechnology* 10: 274–5.

Caseley, J. C., G. W. Cussans and R. K. Atkins (1991) *Herbicide-Resistant Weeds and Crops*, Butterworth-Heinemann, Oxford.

Chyi, Y.-S., R. A. Jorgensen, D. Goldstein, S. D. Tanksley and F. Loaiza-Figueroa (1986) 'Locations and stability of *Agrobacterium*-mediated T-DNA insertions in the *Lycopersicon* genome', *Mol. Gen. Genet.* 204: 64–9.

Comai, L., D. Faccioti, W. R. Hiatt, G. Thompson, R. E. Rose and D. M. Stalker (1985) 'Expression in plants of a mutant aroA gene from *Salmonella typhimurium* confers tolerance to glyphosate', *Nature* 317: 741–4.

Cory, J. S., M. L. Hirst, T. Williams, R. S. Hails, D. Goulson, B. M. Green, T. M. Carty, R. D. Posee, P. J. Cayley and D. K. L. Bishop (1994) 'Field trial of a genetically improved baculovirus', *Nature* 370: 138–40.

Crawley, M. J., R. S. Hails, M. Rees, D. Kohn and J. Buxton (1993) 'Ecology of transgenic oilseed rape in natural habitats', *Nature* 363: 620–3.

Dale, P. J., J. A. Irwin and J. A. Scheffler (1993) 'The experimental and commercial release of transgenic crop plants', *Plant Breeding* 111: 1–22.

Darwin, C. R. (1859) *The Origin of Species by Means of Natural Selection or the Preservation of Favourable Races in the Struggle for Life*, first published by J. W. Burrow, reprinted 1968, Penguin, Harmondsworth.

Davidmann, M. (1996) 'Community Economics: multinational operations. Creating, patenting and marketing new forms of life', http://www.demon.co.uk/solbaram/articles/clm505.html

Dawkins, R. (1976) *The Selfish Gene*, Oxford University Press, Oxford.

Deakin, R. (1990) 'BST: the first commercial product for agriculture from biotechnology', in P. Wheale and R. McNally (eds), *The Bio-Revolution*, Pluto Press, London, pp. 64-73.

De Block, M., J. Botterman, M. Vandewiele, J. Dockx, C. Theon, V. Gosselé, N. R. Movva, C. Thompson, M. Van Montagu and J. Leemans (1987) 'Engineering herbicide resistance in plants by expression of a detoxifying enzyme', *EMBO Journal* 6: 2513–18.

Delannay, X., T. W. Bauman, D. H. Beighley, M. J. Buettner, H. D. Coble, M. S. DeFelice, C. W. Derting, T. J. Diedrick, J. L. Griffen, E. S. Hagood, F. G. Hancock, S. E. Hart, B. J. LaVallee, M. M. Loux, W. E. Lueschen, K. W. Matson, C. K. Moots, E. Murdock, A. D. Nickell, M. D. K. Owen, E. H. Paschall II, L. M. Prochaska, P. J. Raymond, D. B. Reynolds, D. B. Rhodes, F. W. Roeth, P. L. Sprankle, L. J. Tarochione, C. N. Tinius, R. H. Walker, L. M. Wax, H. D. Weigelt and S. R. Padgette (1995) 'Yield evalution of a glyphosphate-tolerant soybean line after treatment with glyphosphate', *Crop Science* 35: 1461–7.

Devlin, R. H., T. Y. Yesaki, C. A. Biagi, E. M. Donaldson, P. Swanson and W. Chan (1994) 'Extraordinary salmon growth', *Nature* 371: 209–10.

Dixon, P. (1995) *The Genetic Revolution*, Kingsway, Eastbourne.

Doyle, J. D., G. Stotzky, G. McClung and C. W. Hendricks (1995) 'Effects of genetically engineered micro-organisms on microbial populations and processes in natural habitats', *Advances in Applied Microbiology* 40: 237–87.

Ellis, R. J. (1983) 'Mobile genes of chloroplasts and the promiscuity of DNA', *Nature* 304: 308–9.

Feitelson, J. S., J. Payne and L. Kim (1992) '*Bacillus thuringiensis*: insects and beyond', *BioTechnology* 10: 271–5.

Fischhoff, D. A., K. S. Bowdish, F. J. Perlack, P. G. Marrone, S. M. McCormick, J. G. Niedermeyer, D. A. Dean, K. Kusano-Kretzmer, E. J. Meyer, D. E. Rochester, S. G. Rogers and R. T. Fraley (1987) 'Insect tolerant transgenic tomato plants', *BioTechnology* 5: 807–13.

Fraley, R., S. Roger, R. Horsch, P. Sanders, J. Flick, S. Adams, M. Bittner, L. Brand, C. Fink, J. Fry, G. Galluppi, S. Goldberg, N. Hoffmann and S. Woo (1983) 'Expression of bacterial genes in plant cells', *Proceedings of the National Academy of Sciences, USA*, 80: 4803–7.

Frossard, P. (1991) *The Lottery of Life: The New Genetics and the Future of Mankind*, Transworld, London.

Gatehouse, A. M. R. and D. Boulter (1983) 'Assessment of antimetabolic effects of trypsin inhibitors from cowpea (*Vigna unguiculata*) and other legumes on development of the bruchid beetle *Callosobruchus maculatus*', *Journal of the Science of Food and Agriculture* 34: 345–50.

Gatehouse, J. A., D. Bown, I. M. Evans, L. N. Gatehouse, D. Jobes, P. Preston and R. R. D. Croy (1987) 'Sequence of the seed lectin gene from pea (*Pisum sativum* L.)', *Nucleic Acids Research* 15: 7642.

Goldsmith, E. (1990) 'The Uruguay Round: gunboat diplomacy by another name', *The Ecologist* 20 (6): 202–4.

Goodman, S. F. (1993) *The European Community*, 2nd edn, Macmillan, London.

Goy, P. A. and J. H. Duesing (1995) 'From pots to plots: genetically modified plants on trial', *BioTechnology* 13: 454–8.

Greene, A. E. and R. F. Allison (1994) 'Recombination between viral RNA and transgenic plant transcripts', *Science* 263: 1423–5.

Gressel, J. (1993) 'Advances in achieving the needs for biotechnologically-derived herbicide-resistant crops', *Plant Breeding Reviews* 11: 155–98.

Grierson, D. (1996) 'Silent genes and everlasting fruits and vegetables', *Nature Biotechnology* 14: 828–9.

Grierson, D. and S. N. Covey (1988) *Plant Molecular Biology*, 2nd edn, Blackie, London.

Grierson, D., M. J. Maunders, M. J. Holdsworth, J. Ray, C. Bird, P. Moureau, W. Schuch, A. Slater, J. E. Knapp and G. A. Tucker (1987) 'Expression and function of ripening genes', in D. J. Nevins and R. A. Jones (eds), *Tomato Biotechnology*, Alan R. Liss, New York, pp. 309-23.

Grossmann, E. and D. Atkinson (eds) (1985) *The Herbicide Glyphosate*, Butterworth, London.

Hammond, B. G., J. L. Vicini, G. F. Hartnell, M. W. Naylor, C. D. Knight, E. H. Robinson, R. L. Fuchs and S. R. Padgette (1996) 'The feeding value of soybeans fed to rats, chickens, catfish, and dairy cattle is not altered by genetic incorporation of glyphosate tolerance', *Journal of Nutrition* 126: 717–27.

Harrison, L. E., M. R. Bailey, M. W. Naylor, J. E. Ream, B. G. Hammond, D. L. Nida, B. L. Burnette, T. E. Nickson, T. A. Mitsky, M. L. Taylor, R. L. Fuchs and S. R. Padgette (1996) 'The expressed protein in glyphosate-tolerant soybean, 5-enolpyruvylshikimate-3-phosphate synthase from *Agrobacterium* sp. strain CP4, is rapidly digested *in vitro* and is not toxic to acutely gavaged mice', *Journal of Nutrition* 126: 728–40.

Hiatt, A., R. Cafferkey and K. Bowdish (1989) 'Production of antibodies in transgenic plants', *Nature* 342: 76–82.

Hightower, R., C. Baden, E. Penzes, P. Lund and P. Dunsmuir (1991) 'Expression of antifreeze proteins in transgenic plants', *Plant Molecular Biology* 17: 1013–21.

Hilder, V. A., A. M. R. Gatehouse, S. E. Sheerman, R. F. Barker and D. Boulter

(1987) 'A novel mechanism of insect resistance engineered into tobacco', *Nature* 330: 160–3.

Hilder, V. A., A. M. R. Gatehouse and D. Boulter (1993) 'Transgenic plants conferring insect resistance: protease inhibitors', in S. D. Kung and R. Wu (eds), *Transgenic Plants Vol 1: Engineering and Utilization*, Academic Press, New York, pp. 317-38.

Hinchee, M. A. W., D. V. Connor-Ward, C. A. Newell, R. E. McDonnell, S. J. Sato, C. S. Gasser, D. A. Fischhoff, D. B. Re, R. T. Fraley and R. B. Horsch (1988) 'Production of transgenic soybean plants using *Agrobacterium*-mediated gene transfer', *BioTechnology* 6: 915–22.

Hinchee, M. A. W., S. R. Padgette, G. M. Kishore, X. Delannay and R. T. Fraley (1993) 'Herbicide-Tolerant Crops', in S. D. Kung and R. Wu (eds), *Transgenic Plants Vol 1: Engineering and Utilization*, Academic Press, New York, pp. 243-63.

Ho, M. W. (1996) 'Perils amid promises of genetically modified foods', Open University, UK, http://www.greenpeace.org/~comms/cbio/geperil.html

Hobbelink, H. (1991) *Biotechnology and the Future of World Agriculture*, Zed Books, London.

Hoffmann, L. M., D. D. Donaldson, R. Brookland, K. Rashka and E. M. Herman (1987) 'Synthesis and protein body deposition of maize 15-Kd zein in transgenic tobacco seed', *EMBO Journal* 6: 3213–21.

Hoffmann, T., C. Golz and O. Schieder (1994) 'Foreign DNA sequences are received by a wild-type strain of *Aspergillus niger* after co-culture with transgenic higher plants', *Current Genetics* 27: 70–6.

Iyengar, A., F. Müller and N. MacLean (1996) 'Regulation and expression of transgenes in fish – a review', *Transgenic Research* 5: 147–66.

Jenes, B., H. Moore, J. Cao, W. Zhang and R. Wu (1993) 'Techniques for gene transfer', in S. D. Kung and R. Wu (eds), *Transgenic Plants Vol 1: Engineering and Utilization,* Academic Press, USA, pp. 125–46.

Joergensen, R. B. and B. Andersen (1994) 'Spontaneous hybridization between oilseed rape, *Brassica napus*, and weedy *B. Campestris* (Brassicaceae): a risk of growing genetically modified oilseed rape' *American Journal of Botany* 81: 1620–6.

Juma, C. (1989) *The Gene Hunters: Biotechnology and the Scramble for Seeds*, Zed Books, London.

Juskevich, J. C. and C. G. Guyer (1990) 'Bovine growth hormone: human food safety evaluation', *Science* 249: 875–83.

Kennedy, P. (1993) *Preparing for the Twenty-First Century,* HarperCollins, London.

Klein, T. M., E. D. Wolf, R. Wu and J. C. Sanford (1987) 'High-velocity microprojectiles for delivering nucleic acids into living cells', *Nature* 327: 70–3.

Klinger, T. and N. C. Ellstrand (1994) 'Engineered genes in wild populations: fitness of weed-crop hybrids of *Raphanus sativus*, *Ecological Applications* 4: 117–20.

Klinger, T., P. E. Arriola and N. C. Ellstrand (1992) 'Crop-weed hybridization in radish (*Raphanus sativus* L.): effects of distance and population size', *American Journal of Botany* 79: 1431–5.

Kung, S. D. and R. Wu (1993a) *Transgenic Plants Vol 1: Engineering and Utilization*, Academic Press, New York.

Kung, S. D. and R. Wu (1993b) *Transgenic Plants Vol 2: Present Status and Social and Economic Impacts*, Academic Press, USA.

Lal, R. and S. Lal (1990) *Crop Improvement Utilizing Biotechnology*, CRC Press, Boca Raton, FL.

Landsmann, J. and A. Shah (1995) 'Biosearch: Genetechnik-Detenbank der BBA, 3', Mitteilung, *Nachrichtenblatt des Deutschen Pflanzenschutzdienstes* 47 (5): 135.

Landsmann, J., A. Shah and R. Casper (1995) 'Biosearch: Genetechnik-Detenbank der BBA, 3', Mitteilung, *Nachrichtenblatt des Deutschen Pflanzenschutzdienstes* 47 (4): 102.

Lindow, S. E. and N. J. Panopoulos (1988) 'Field tests of recombinant ice minus *Pseudomonas syringae* for biological frost control in potato', in M. Sussman, C. H. Collins, F. A. Skinner and D. E. Stewart-Tull (eds), *Release of Genetically-engineered Micro-organisms,* Academic Press, New York.

Lycett, G. W. and D. Grierson (1990) *Genetic Engineering of Crop Plants*, Butterworth, London.

McCabe, D. E., W. F. Swain, B. J. Martinell and P. Christou (1988) 'Stable transformation of soybean (*Glycine max*) by particle acceleration', *BioTechnology* 6: 923–6.

McClintock, B. (1951) 'Chromosome organisation and genic expression' *Cold Spring Harbour Symposium Quant. Biol.* 16: 13–47.

Maiti, I. B., G. J. Wagoner, R. Yeargan and A. G. Hunt (1989) 'Inheritence and expression of the mouse metallothionein gene in tobacco', *Plant Physiology* 91: 1020–4.

Meeusen, R. L. and G. Warren (1989) 'Insect Control with genetically engineered crops', *Annual Review of Entomology* 34: 373–81.

Miki, B. I., H. Labbe, J. Hattori, J. Ouellet, J. Gabard, G. Sunohara, P. J. Charest and V. N. Iyer (1990) 'Transformation of *Brassica napus* canola cultivars with *Arabidopsis thaliana* acetohydroxyacid synthase genes and analysis of herbicide resistance', *Theoretical and Applied Genetics* 80: 449–58.

Miller, H. (1994) 'Risk assessment experiments and the new biotechnology', *Trends in Biotechnology* 12: 292–5.

Miller, H. (1997) *Policy Controversy in Biotechnology: An Insider's View*, RG Landes, Austin, TX.

Millstone, E. (1986) *Food Additives: Taking the Lid Off What We Really Eat*, Penguin, Harmondsworth.

Murata, N., O. Ishizaki-Nishizawa, S. Higashi, H. Hoyashi, Y. Tasaka and I. Nishida (1992) 'Genetically engineered alterations in the chilling sensitivity of plants', *Nature* 356: 710–13.

Nap, J. P., J. Bijvoet and W. J. Stiekema (1992) 'Biosafety of kanamycin-resistant transgenic plants', *Transgenic Research* 1: 239–49.

Nestle, M. (1996) 'Allergies to transgenic foods – question of policy', *New England Journal of Medicine* 334 (11): 726–8.

Nordlee, J. A., S. L. Taylor, J. A. Townsend, L. A. Thomas and R. K. Bush (1996) 'Identification of a brazil-nut allergen in transgenic soybeans', *New England Journal of Medicine* 334 (11): 688–92.

OECD (Organization for Economic Cooperation and Development) (1986) *Recombinant DNA Safety Considerations*, OECD, Paris.

Oka, I. N. and D. Pimentel (1976) 'Herbicide (2,4–D) increased insect and pathogen pests on corn', *Science* 193: 239.

OTA (Office of Technology Assessment) (1987) *New Developments in Biotechnology – Background Paper: Public Perceptions of Biotechnology*, OTA, US Government Printing Office, Washington, DC.

Ow, D .W., K. V. Wood, M. DeLuca, J. R. DeWet, D. R. Helinski and S. H. Howell (1986) 'Transient and stable expression of the firefly gene in plant cells and transgenic plants', *Science* 234: 856–9.

Oxtoby, E. and M. A. Hughes (1990) 'Engineering herbicide tolerance into crops', *Trends in Biotechnology* 8: 61–5.

Padgette, S. R., N. B. Taylor, D. L. Nida, M. R. Bailey, J. MacDonald, L. R. Holden and R. L. Fuchs (1996) 'The composition of glyphosate-tolerant soybean seeds is equivalent to that of conventional soybeans', *Journal of Nutrition* 126: 702–16.

PANOS (1995) 'Pharmaceuticals target northern seed collections', Media Briefing no. 15, D. De Sarkar, http://www.oneworld.org/panos/oct96/seed.html

Pear, J. R., R. A. Sanders, K. R. Summerfelt, B. Martineau and W. Hiatt (1993) 'Simultaneous inhibition of two tomato cell wall hydrolases, pectinmethylesterase and polygalacturonase, with antisense constructs', *Antisense Research and Development* 3: 181–90.

Penarrubia, L., R. Kim, J. Giovannoni, S.-H. Kim and R. L. Fischer (1992) 'Transgenic expression of monellin in tomato and lettuce', *BioTechnology* 10: 561–4.

Penman, D. (1996) *The Price of Meat: Salmonella, Listeria, Mad Cows – What Next?* Gollancz, London.

Piccioni, R. (1990) 'Food irradiation: contaminating our food', *The Ecologist* 18 (2): 48–55.

Pollack, R. (1994) *Signs of Life: The Language and Meanings of DNA*, Viking Penguin, Harmondsworth.

Poirier, Y., D. E. Dennis, K. Klomparens and C. Somerville (1992) *Polyhydroxybutyrate, a biodegradable plastic, produced in transgenic plants*, *Science* 256: 520–3.

Potrykus, I. (1990) 'Gene transfer to cereals: an assessment', *BioTechnology* 8: 535–42.

Pursel, V. G., C. E. Rexroad Jr., D. J. Bolt, K. F. Miller, R. J. Wall, R. E. Hammer, C. A. Pinkert, R. D. Palmiter and R. L. Brinster (1987) Progress on gene transfer in farm animals', *Veterinary Immunology and Immunopathology* 17: 303–12.

Quinn, J. P. (1990) 'Evolving strategies for the genetic engineering of herbicide resistance in plants', *Biotechnological Advances* 8: 321–33.

Rabinow, P. (1996) *Making PCR: A Story of Biotechnology*, University of Chicago Press, Chicago.

RAFI (Rural Advancement Foundation International) (1989) 'Vanilla and biotechnology', *RAFI Communiqué*, http://www.rafi.ca/communique/19891.html

RAFI (1993) 'Biotechnology company will sell bio-engineered human proteins to infant formula manufacturer', *RAFI Communiqué*, http://www.rafi.ca/communique/19933.html

RAFI (1994) '"Species" patent on transgenic soybeans granted to transnational chemical giant W.R. Grace', *RAFI Communiqué*, http://www.rafi.ca/communique/19942.html

RAFI (1995) 'Genetically engineered high-lauric rapeseed (canola): what threat to tropical lauric acid production', *RAFI Communiqué*, http://www.rafi.ca/communique/19952.html

RAFI (1997) 'Bioserfdom: technology, intellectual property and the erosion of farmers' rights in the industrialized world', *RAFI Communiqué*, http://www.rafi.ca/communique/19972.html

Raghavan, C. (1990) *Recolonisation: GATT, the Uruguay Round and the Third World*, Third World Network, Penang, Malaysia.

Rhodes, C. A., D. A. Pierce, I. J. Mettler, D. Mascarenhas and J. J. Detmer (1988) 'Genetically transformed maize plants from protoplasts', *Science* 240: 204–7.

Rissler, J. and M. Mellon (1996) *The Ecological Risks of Engineered Crops*, MIT Press, Cambridge, MA.

Robert, L. S., R. D. Thompson and R. B. Flavell (1989) 'Tissue-specific expression of a wheat high molecular weight glutein gene in transgenic tobacco', *Plant Cell* 1: 569–78.

Royal Commission on Environmental Pollution (1989) *The Release of Genetically Engineered Organisms to the Environment*, Thirteenth Report, HMSO, London.

Ryan, C. A. (1990) 'Protease inhibitors in plants: genes for improving defences against insects and pathogens', *Annual Review of Phytopathology* 23: 425–49.

Sachs, E. S., J. H. Benedict, J. F. Taylor, D. M. Stelly, S. K. Davis and D. W. Altman (1997) 'Pyramiding CryIA(b) insecticidal protein and terpenoids in cotton to resist tobacco budworm (Lepidoptera: Noctuidae)', *Environmental Entomology* 25: 1257–66.

Schmutterer, H. and K. R. S. Ascher (eds) (1984) *Natural Pesticides from the Neem Tree and Other Tropical Plants*, Schriftenreihe der GTZ, no. 161, Eschborn.

Schöpke, C., N. Taylor, R. Cárcamo, N'K. Konan, P. Marmey, G. G. Henshaw, R. N. Beachy and C. Fauquet (1996) 'Regeneration of transgenic cassava (*Manihot esculenta* Crantz) from microbombarded embryogenic suspension cultures', *Nature Biotechnology* 14: 731–40.

Skogsmyr, I. (1994) 'Gene dispersal from transgenic potatoes to conspecifics: a field trial', *Theoretical and Applied Genetics* 88: 770–4.

Stewart, L .M. D., M. Hirst, M. López Ferber, A. T. Merryweather, P. J. Cayley and R. D. Possee (1991) 'Construction of an improved baculovirus insecticide containing an insect-specific toxin gene', *Nature* 352: 85–8.

Summerfield, R. J. and E. H. Roberts (1983) 'The soyabean' *Biologist* 30 (4): 223–31.

Tabashnik, B. E. (1994) 'Evolution of resistance to *Bacillus thuringiensis*', *Annual Review of Entomology* 39: 47–9.

Tabashnik, B. E., Y.-B. Liu, N. Finson, L. Masson and D. G. Heckel (1997) 'One gene in diamondback moth confers resistance to four B.t. toxins', *Proceedings of the National Academy of Sciences, USA*, 94: 1640–4.

Tate, J. (1990) 'NIMBY and NIAMBY: public perceptions of biotechnology', in P. Wheale and R. McNally (eds), *The Bio-Revolution*, Pluto Press, London, pp. 224-36.

Tepfer, M. and M. Jacquemond (1996) 'Sleeping satellites: a risky prospect', *Nature Biotechnology* 14: 1226.

Timberlake, L. (1985) *Africa in Crisis: The Causes, the Cures of Environmental Bankruptcy*, Earthscan, London.

Todaro, M. P. (1989) *Economic Development in the Third World*, 4th edn, Longman, New York.

Tomalski, M. D. and L. K. Miller (1991) 'Insect paralysis by baculovirus-mediated expression of a mite neurotoxin gene', *Nature* 352: 82–5.

Townsend, J. A., L. A. Thomas, E. S. Kulisek, M. J. Daywalt, K. R. K. Winter, D. J. Grace, W. J. Crook, H. J. Schmidt, T. C. Corbin and S. B. Altenbach (1992) 'Accumulation of a methionine-rich brazil nut protein in seeds of transgenic soyabean', *American Society of Agronomy. Abstract* 198.

Tudge, C. (1988) *Food Crops for the Future*, Blackwell, Oxford.

Tudge, C. (1993) *The Engineer in the Garden. Genetics: From the Idea of Heredity to the Creation of Life*, Jonathan Cape, London.

Vaeck, M., A. Reynaerts, H. Höfte, S. Jansens, M. De Beuckeleer, C. Dean, M. Zabeau, M. Van Montagu and J. Leemans (1987) 'Transgenic plants protected from insect attack', *Nature* 328: 33–7.

Vavilov, N. I. (1951) *The Origin, Variation, Immunity and Breeding of Cultivated Plants*, Roland Press, New York.

Voelker, T. A., A. C. Worrell, L. Anderson, J. Bleibaum, C. Fan, D. J. Hawkins, S. E. Radke and H. M. Davies (1992) 'Fatty acid biosynthesis redirected to medium chains in transgenic plants', *Science* 257: 72–4.

Voelker, T. A., T. R. Hayes, A. M. Cranmer, J. C. Turner and H.M. Davies (1996) 'Genetic engineering of a quantitative trait: metabolic and genetic parameters influencing the accumulation of laurate in rapeseed', *Plant Journal* 9: 229–41.

Watson, J. D. (1968) *The Double Helix: A Personal Account of the Discovery of the Structure of DNA*, Penguin, Harmondsworth.

Webb, T. and T. Lang (1990) *Food Irradiation: The Myths and the Reality*, Thorsons, London.

Webb, V. and J. Davies (1994) 'Accidental release of antibiotic-resistance genes', *Trends in Biotechnology* 12: 74–5.

Wheale, P. and R. McNally (eds) (1990) *The Bio-Revolution*, Pluto Press, London.

White, F. F. (1993) 'Vectors for gene transfer in higher plants', in S. D. Kung and R. Wu (eds), *Transgenic Plants Vol 1: Engineering and Utilization*, Academic Press, New York, pp. 15–48.

Wilmut, I., A. E. Schieke, J. McWhir, A. J. Kind and K. H. S. Campbell (1997) 'Viable offspring derived from fetal and adult mammalian cells', *Nature* 385: 810–13.

Wood, H. A. (1995) 'Development and testing of genetically improved baculovirus insecticides', in M. L. Shuler, H. A. Wood, R. R. Granados and D. A. Hammer (eds), *Baculovirus Expression Systems and Biopesticides*, Wiley, New York, pp. 91–108.

Index